T0220249

Backseat Driver

Buying the safest car for your family shouldn't be up for debate.

Yet for decades, car safety advocates, manufacturers, and lawmakers in the United States have clashed over whether to make automobiles safer. All sides armed themselves with data in the hopes of winning the great car safety debates. In this way, crash statistics and the analysts who studied them made history. But data were always in the backseat, merely supporting different points of view. That is, until now.

With car safety, it's the value we place on every human life that counts.

Automobile safety expert Dr. Norma Faris Hubele delivers a lively discussion of the role data play in protecting you and your family on the road. You'll gain a greater appreciation for how:

- A World War I pilot's near-death experience birthed the U.S. car safety movement
- Data from real car crashes helped create the first vehicle safety standards
- A shift toward fuel-efficient cars affected fatality risk in the 1970s–1980s versus now
- Vehicle size has changed, and the problems that creates for you and others sharing the road
- Car safety rating systems, even when limited, empower consumers and motivate manufacturers
- Federal regulators decide whether to issue a safety recall on your vehicle
- Data's role is evolving with the advent of driver-assist and self-driving technologies

Norma Faris Hubele, PhD has served over 30 years as a professor, consumer advocate, and automotive safety expert. She was the first director of strategic initiatives at the Fulton School of Engineering at Arizona State University, where she is now a Professor Emerita. An expert witness in over 120 car safety cases, Dr. Hubele has educated the courts about when statistics can or cannot inform decisions. She is a co-author of a widely used textbook, *Engineering Statistics*. Dr. Hubele is founder and CEO of TheAutoProfessor.com, a free website for new car safety ratings and information.

ASA-CRC Series on Statistical Reasoning in Science and Society

Series Editors
Nicholas Fisher, University of Sydney, Australia
Nicholas Horton, Amherst College, MA, USA
Regina Nuzzo, Gallaudet University, Washington, DC, USA
David J Spiegelhalter, University of Cambridge, UK

Published Titles

Errors, Blunders, and Lies: How to Tell the Difference
David S. Salsburg

Visualizing Baseball
Jim Albert

Data Visualization: Charts, Maps and Interactive Graphics
Robert Grant

Improving Your NCAA® Bracket with Statistics
Tom Adams

Statistics and Health Care Fraud: How to Save Billions
Tahir Ekin

Measuring Crime: Behind the Statistics
Sharon Lohr

Measuring Society
Chaitra H. Nagaraja

Monitoring the Health of Populations by Tracking Disease Outbreaks
Steven E. Fricker and Ronald D. Fricker, Jr.

Debunking Seven Terrorism Myths Using Statistics
Andre Python

Achieving Product Reliability: A Key to Business Success
Necip Doganaksoy, William Q. Meeker, and Gerald J. Hahn

Protecting Your Privacy in a Data-Driven World
Claire McKay Bowen

Backseat Driver: The Role of Data in Great Car Safety Debates
Norma Faris Hubele

For more information about this series, please visit: https://www.routledge.com/go/asacrc

Backseat Driver
The Role of Data in
Great Car Safety Debates

Norma Faris Hubele

Founder and CEO of TheAutoProfessor.com
Professor Emerita at Arizona State University

CRC Press
Taylor & Francis Group
Boca Raton London New York

CRC Press is an imprint of the
Taylor & Francis Group, an **informa** business
A CHAPMAN & HALL BOOK

In substance the Work contains the author's opinion based on safety standards and her experience and no liability of the author, the publisher, or any other third party will ensue for a reader's reliance on or lack of a reliance on the information in the Work.

First Edition published 2023
by CRC Press
6000 Broken Sound Parkway NW, Suite 300, Boca Raton, FL 33487-2742

and by CRC Press
4 Park Square, Milton Park, Abingdon, Oxon, OX14 4RN

CRC Press is an imprint of Taylor & Francis Group, LLC

ISBN: 978-0-367-47407-2 (hbk)
ISBN: 978-0-367-47230-6 (pbk)
ISBN: 978-1-003-03534-3 (ebk)

DOI: 10.1201/9781003035343

To Norman

Contents

CHAPTER 1

Introduction

In closing, let me emphasize that more than 100 people die on our highways every single day. In our view, one death is too many. We must change a culture that is willing to accept those losses, and we need your help to implement proven solutions.

—*JENNIFER HOMENDY (2019)*[1]

Tommy, 4 years old, died in a fiery car crash in March of 2012. He was secure in a child safety seat in the back of his family's 1999 Jeep Grand Cherokee. His older cousin, Sally, was driving him to tennis class when a driver, high on cocaine, plowed into the rear of their SUV.

The fire, fueled by the Jeep's rear-mounted gas tank, spread so fast and violently that nobody could help Tommy escape. Sally was uninjured, as was the driver who caused the crash. If not for the fire, Tommy only would have suffered a fractured leg.

Fourteen years before this crash, a mother wrote a prophetic letter to Jeep's corporate offices about her daughter's harrowing escape from one of their vehicles.

Within moments the Jeep was on fire because the gas tank had been hit. The driver's door was already in flames when she jumped out of the car. However, in thinking about it afterwards, I can only imagine how horrible a situation it would be if a driver had

DOI: 10.1201/9781003035343-1

to remove a child from a car seat or could not get out of the car within moments.

The mother was among many who voiced their concern about Jeep's design decision to place the gas tank behind the rear axle, where it is vulnerable in rear-impact crashes. Comparisons were drawn between the Grand Cherokee and the infamous Ford Pinto with its rear-mounted gas tank, which caused violent passenger deaths and led to a recall in the 1970s.

At the time of Tommy's death, the 1999 Jeep Grand Cherokee was under scrutiny by the Office of Defects Investigation (ODI) at the National Highway Traffic Safety Administration (NHTSA). This office handles vehicle recalls related to safety. They were investigating fuel tank failures and vehicle fires in rear impacts. The ODI had studied fatal crash statistics and determined that Jeep vehicles had a higher incidence of fatalities in these types of crashes than other similar size vehicles.

On June 12, 2012, the ODI raised their investigation to the status of an Engineering Analysis. Three types of Jeeps with rear-mounted fuel talks were edging closer to a recall: Jeep Grand Cherokee (model years 1993–2004), the Jeep Cherokee (1993–2001), and the Jeep Liberty (2002–2007). The ODI asked the Chrysler Group – Jeep's corporate organization – to respond.

Analysts working for Chrysler challenged the ODI's findings. Using different statistics and comparison vehicles, these analysts argued that the Jeeps had better overall safety records than other comparison vehicles. In fact, their vehicles met or exceeded all applicable Federal Motor Vehicle Safety Standards. In addition, there were plenty of other vehicles, past and present, that had worse fire rates and had never faced a recall. And notably, the Jeep fires occurred in severe, high-energy rear impacts.

Personal tragedy spurs car safety debates. Tommy's death was one of those tragedies. That's where the data come in. The data are the "counts." How many deaths? How many injuries? How severe? To some decision-makers, the data show a problem. Others argue that there are miscounts. A different way of counting reveals no problem.

Numerous factors influence decisions in car safety. The stakes in lives, money, political capital, and brand reputation are very high. It's not surprising that in the public arena, each side seeks to use data to make their case.

Ultimately, a decision is made. The ODI issues a recall. A jury issues a verdict. A carmaker changes a vehicle design. Did the data make the difference? Or, was it the bad publicity? The high cost of law suits? Changes in customer preferences? Or, was it just the right combination of all of the above? In truth, we never know for certain what tilted the scales.

The ODI ended its investigation into the Jeep-brand vehicles on November 14, 2014. In doing so, it accepted Chrysler's proposal to install a hitch receiver, or tow hitch, on 1993–1998 Jeep Grand Cherokee and 2002–2007 Jeep Liberty vehicles. Laboratory testing of rear impacts with these hitches provided enough evidence for the ODI to conclude that the remedy had "incremental safety benefits in certain low- and moderate-speed crash incidents." Chrysler's data analysis even convinced the ODI to exclude Tommy's 1999 Grand Cherokee in the recall. In the document ending their investigation, the ODI wrote:

> Examination of the available data established that the MY [model years] 1999–2004 Grand Cherokee did not pose the same magnitude of safety risk as the MY 1993–1998 Grand Cherokee and MY 2002–2007 Liberty, particularly in low- and moderate-speed rear impacts.

Chrysler no longer builds vehicles with rear-mounted gas tanks.

FRAMING THE DISCUSSION

Since this book is about using data to make our cars safer, it seems natural to start with some simple questions. When did the United States start collecting in-depth data on car crashes? What motivated it? Who did the collecting? What form did it take? How did that collection evolve to the present-day databases of information on millions of crashes?

To answer these questions in detail is beyond the scope of this book. However, I set the stage for our discussion by including a few historical highlights in the following two chapters. You'll see how key people and public attitudes shaped what was happening in the early- and mid-20th century, leading to our present treasure trove of information.

I begin with a look at the early days of the automobile. Road deaths and injuries rose sharply and details about these crashes were lacking. Most people dismissed these tragedies with a "too bad" type of attitude, and the public usually blamed the drivers. The prevailing view was that these crashes were simply the cost of modern transportation.

In the mid-20th century, attitudes started to change. A group of doctors, statisticians, and engineers carefully compiled evidence that showed some injuries and fatalities could be prevented with proper car design. As a society, we might not be able to prevent all crashes, but we could possibly reduce or mitigate injuries in a crash by making vehicles safer.

Safety advocates began to ask key questions about what could be fixed in cars. Today, cars in the United States must meet more than 40 federal standards to ensure safety. These include padded dashboards, collapsing steering columns, seat belts, airbags, and better door locks, just to name a few. Safety experts credit these standards with saving over 620,000 lives since 1960.

But there was much debate leading up to this progress. In the remaining chapters, I describe some data-intensive controversies that accompanied these gains. These examples shine a light on the analyses performed at the NHTSA. This division of our federal Department of Transportation has a mission to save lives, as well as reduce injury and property damage on our roadways.

NHTSA uses data to carry out its mission, whether it is creating new standards, updating old ones, deciding on recalls, or rating the safety of our cars. And so, do its opponents. These data-driven debates are the focus of this book. How, when, and why different datasets and types of analyses are used to determine, ultimately, the type of cars we get to drive.

The data discussed in this book are not complicated. Simple graphs and tables capture the core elements of the disputes. If you want to find the sources for the data in the tables and graphs or you want to go deeper into the discussions, the **Further Reading** for each chapter will provide you with a starting point. However, you don't need a college-level course in statistics to read this book. You only need a curious mind and an appreciation for debating vital issues facing our society.

THE DEBATES

It would take volumes to cover all the great debates in car safety. Instead, I chose a set of prime examples to illustrate data's role in some of the most intense disputes among lawmakers, safety advocates, and carmakers.

The chapters use elements of real-life stories to illustrate a controversial topic in car safety. For example, I discuss Tommy's tragic death and Jeep's argument for the rear-mounted fuel tank in the context of the recall debates in Chapter 9. In another recall dispute, I show how one owner's complaint was part of the investigation of unintended acceleration allegations in Tesla vehicles. In another chapter, a story about a fellow's death in a rollover crash demonstrates how heated the public debates around federal safety standards can be. The back and forth between sides can get tedious, but the main message is clear. Each side is certain that their analysis is the right one. In the end, the federal administrators must make a decision.

Sometimes, the car safety problem discussed in a chapter needs a lot of context. You'll see that in the chapter on the fuel efficiency debate. Here, I describe the Corporate Average Fuel Efficiency (CAFE) standard before I explain the role of data in the debate. Over the past 30 years, this story took some twists and turns. Environmentalists wanted higher targets for miles-per-gallon, while industry leaders wanted less regulation. Passions ran high. As you will see, at each junction in the road, these competitors armed themselves with data to support different points of view.

In the final chapter, I talk about self-driving cars. These high-tech vehicles vow to remove human error from our roadways, thus saving lives, injuries, and property damage. Their makers suggest that these vehicles are all about safety. And where does data play a role? Everywhere. Data are used to create these robots and test them. Data are used to proclaim these futuristic vehicles are ready to hit the road alongside you and me. And data are used to block their deployment as well. Society is moving toward a future in which data are no longer in the backseat advising – the data are behind the steering wheel driving. Carmakers are investing billions of dollars in the promise of these autonomous vehicles. It's no wonder that the debates are so fierce.

In each chapter, I make a concerted effort to give you an objective view of the debate. However, early reviewers of this manuscript thought you also might like to have my personal take. To this end, I added a section called **Note From the Author**. Here, I give you my opinion about what transpired in the debate and my thoughts about the outcome. These opinions are based on my personal knowledge of these car safety issues. They reflect my experience as a statistician for over 30 years; as an expert witness in over 120 car safety cases; as a university professor teaching student engineers; and now, as founder of TheAutoProfessor.com, a free website that uses data to help families choose safer vehicles.

THE VALUE PROPOSITION

This is a book about data and social progress. The subject matter is car safety, but the principles are common across our society. As a democratic, pluralistic country, we are accustomed to debating important issues at the national level. The arguing may be about vaccination rates, immigration, or infrastructure repair. The purpose of these public discussions is to gather information, consider different points of view and, in the end, to reach an informed decision. Today, data are used by all sides.

For many of society's problems, there is no universal way to look at the data. This makes for intense debates. And, what if there are no data or not enough data? Decisions still need to be made. The examples in this book illustrate these situations.

Finally, as a statistician, I am passionate about collecting and analyzing the best data available. But I am also committed to never losing sight of what's behind the data. I humbly use elements of real crashes, such as Tommy's, to remind us that human tragedy is always behind the numbers when talking about any social problem.

My hope is that when you finish reading this book, you will have a new appreciation for the role of data in safeguarding the lives of your family, friends, and neighbors. When it comes to car safety, in particular, and social progress, in general, it's the value we place on every human life that counts.

NOTE

1. Homendy, Jennifer. (2019). *Every life counts: Improving the safety of our nation's roadways, U.S. House of Representatives subcommittee on Highways and Transit,* 116th Cong. (2019) (testimony). https://www.govinfo.gov/content/pkg/CHRG-116hhrg36978/html/CHRG-116hhrg36978.htm. At the time Ms. Homendy gave this testimony, she was a member of the National Transportation Safety Board. On August 13, 2021, she became chairperson.

A Strange Start for a Movement

What is the human being but a lump of mush – protoplasm – sixty
percent water … What a contrast to the car itself. When you suddenly stop a rapidly moving car, so that the protoplasmic passenger,
obeying the second law of Newton, continues on, his protoplasm
spatters. The problem, then, is to protect mush against sudden
impact.

—*EDITORIAL (1937)*[1]

Hugh DeHaven didn't expect a lucky break. He was falling 500 feet
in a plane that was spinning and approaching the ground very fast. He
expected to die.

It was 1917, on the last day of DeHaven's flight training in the Canadian
Royal Flying Corps. As a kid from Brooklyn, DeHaven had crossed into
Canada and volunteered after the U.S. Army Air Corps rejected him a year
earlier. Now he was close to his dream of becoming a pilot. But a reckless
young trainee interfered by deciding to practice gunnery on the tail of
DeHaven's plane. The result was a mid-air collision.

DeHaven lost part of his plane's wing and tail. No recovering from a
fatal blow. Both planes crashed into the ground. DeHaven was the only
one to survive.

His colleagues called him lucky. His superior officer called it the "Jesus
factor." DeHaven, who had a degree in mechanical engineering, never

accepted any of these explanations. He had six months in recovery to think about what happened. His worst injury in the crash came from the seat belt buckle. DeHaven said the force from the crash threw his body forward and onto the stiletto-shaped buckle, which ruptured his pancreas. He felt confident that if he had escaped the abdominal injury, his hospital stay would have been only about one month.

After his recovery, DeHaven worked as a clerk collecting the bodies of deceased pilots after they crashed. That experience added to his observations about the causes of injuries. Some of these crashes looked survivable, if not for the pilot's head hitting the instrument panel or the wall of the cockpit. Figure 2.1 illustrates the type of control panel that DeHaven would have critiqued. He studied how these tragic endings could have been avoided. DeHaven didn't know it at the time, but he was laying the foundation for a new field: the epidemiology of vehicle accidents. And one day, he'd be known as the Father of Crashworthiness Research.

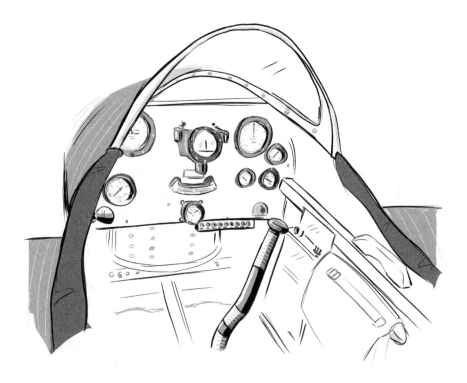

FIGURE 2.1 Illustration of an airplane cockpit circa 1917. DeHaven recognized the protruding dials and sharp edges as safety hazards in a crash.

A REALIZATION

Nearly 20 years after his plane crash, DeHaven witnessed a car crash back in the United States. The driver lost control when his car skidded on a wet road, flipped over, and landed in a ditch. He was not speeding.

The unrestrained driver flew forward and to the right. His head hit the knob for operating the windshield wiper. The sharp steel knob gouged through his right sinus, across his nose, and into the area above his left eye. The injuries should have been minor, but instead they were dangerous, leaving the driver disfigured and disabled.

Recalling this event as a turning point in his life, DeHaven wrote:

> I again realized that engineers didn't know – and that nobody knew – how many times people were hurt or killed by things that could be easily changed. I had seen identical injuries in airplanes and I naturally wondered whether 10 or 100 or 1,000 people had been thrown against similar knobs in automobiles, and whether using a rubber knob would have eliminated their injuries.

Still working among pilots and engineers, DeHaven advocated to improve the safety of planes. Older pilots rebuffed him with statements such as, "If you can't face the dangers of flying, then stay on the ground." Yes, they recognized that the pointed buckle on the pilot's safety belt was quite dangerous. But the belt was only meant to hold a pilot in place during inverted maneuvers. Any smart pilot, if he had time before a crash, would have taken the belt off.

Others mocked his ideas for building a safer cockpit. They'd say, "Sure, we'd be safer in accidents if we beefed up structures and lined the cockpit with mattresses. But that would cost weight and money. We could make planes safer, but we'd never get them off the ground."

DeHaven needed good, reliable data that would not be dismissed. If a pilot survived, he had to be able to clearly show the scientific reasons why. The data had to be undisputable.

With his engineering background, DeHaven understood the idea behind forces on falling objects. A pilot and plane falling from the sky would be subject to these forces. However, these were complicated events with missing data. How high was the plane at the start of the descent? Which forces acted on the plane during the descent and to what degree? What was the pilot's movement within the cockpit during the crash? Even if he could generate estimates, DeHaven knew that his data would be hotly debated and probably rejected.

He conducted a simple experiment. DeHaven dropped eggs from the rungs of a ladder onto a layer of padding on his kitchen floor. As he repeated his experiment, each time from a higher rung, the eggs never broke. Today, what seems like a typical high school physics experiment was bizarre behavior in 1936. DeHaven jokingly said that his wife thought that he might be going insane. But he convinced her that "people knew more about protecting eggs in transit than they did about protecting human heads."

Data from his egg-dropping experiment were interesting, but still far removed from a pilot's experience. DeHaven needed more realistic data. He needed data that directly addressed the capacity of a person's body to withstand falls from tall heights, like a plane crash.

AN IMPORTANT SCIENTIFIC FINDING

Seeing patterns where others didn't, DeHaven began studying failed suicide attempts. He focused on people who jumped from buildings as tall as 50–150 feet. He saw a direct parallel between the survival of pilots in severe plane crashes (similar to his own) and the unlikely survival of these despairing individuals.

DeHaven typically learned of these miraculous events from newspaper articles. The dramatic stories sold newspapers. But DeHaven had a more scientific interest: useful data. He immediately sent an engineer to take measurements and collect data from the sites. What was the height of the fall? What did the person land on? How did the person land? Head first? On her/his back? Was there anything unusual about the landing surface? Did something impede the fall? Did the person get injured? What kind of injuries?

DeHaven believed that he could apply his mechanical engineering knowledge to these measurements and make some calculations. He could compute the forces of acceleration and deceleration on the falling bodies landing on specific surfaces. These measurements and his calculations would become his undisputable dataset.

Each case examined by DeHaven started with a very personal story of despair. However, one story stood out for its thoughtful suicide note.

> Before jumping from a 10th-story window, the 21-year-old woman had built up her courage by drinking a half bottle of whiskey. Three people saw the woman hit the ground. She immediately tried to stand up. The observers cautioned her not to move. The hospital later said that her only injuries were a broken rib and wrist on her right side. She had experienced neither a concussion nor loss of consciousness.

Upon reading about this young woman's survival, DeHaven sent an engineer to the scene. The woman – who was 5 feet 7 inches tall and weighed 115 pounds – fell 93 feet. Her body landed onto a freshly tilled garden in a nearly horizontal position on her right side and back, with the back of her head striking soft earth. DeHaven estimated the forces on the woman's body by combining the elements of the fall and the impressions on the garden bed. The dents in the earth were uneven across the area where the body had landed. The highest force resulted in a 6-inch dent. When the young woman's prone body met the earth with a velocity of 50 miles per hour, the soft earth "gave" 6 inches of cushion.

DeHaven favored this particular case because the data were very clear. It helped to build his hypothesis about the conditions for surviving plane and, in turn, car crashes. Moreover, he was carrying out the woman's wishes. Her suicide note requested that, upon her death, her body should go to science. In a strange twist of fate, it was her survival that "went to science." She had escaped certain death and the reasons why would forever inform the science behind reducing injuries and death in crashes.

DeHaven collected data from seven similar cases to determine the conditions for survival. He concluded that his data supported well-known mechanical and physical laws about injuries dating back to Hippocrates in 400 B.C. The primary causes of injury in crashes were the force of impact and how focused that force was on the body.

To prevent injuries, he suggested distributing the distance (time) and area (space) of the forces on the body. In other words, slow the forces down and blunt their impact. By properly designing cockpits, car compartments, and restraint systems, these engineering principles could save lives. He published his findings in the journal *War Medicine* in 1942.

A GROUP EFFORT

In that same year, DeHaven joined a group of like-minded researchers at Cornell University. Together, they obtained funding to help with the war effort from two federal organizations, the Office of Scientific Research and Development and the National Research Council. The objective of their research was to save lives and money for the armed forces. The problem was the mounting cost of injuries among highly specialized air crew involved in plane accidents. Was this the inevitable price of speed or was there a way to prevent the injuries?

Based on the findings of the Cornell group, the U.S. Air Force learned that their people could be "packaged" in the aircraft, similar to the way fragile merchandise was packaged to avoid damage. They used well-designed safety belts to keep the crew in their seats and to distribute forces on their bodies upon impact. In addition, they redesigned the cockpit, padding surfaces and removing sharp knobs. Injuries dropped dramatically.

In late 1951, a military study revealed that more personnel were hurt in motor vehicle accidents and more time was lost in hospital stays than as a result of the Korean War. Why not apply the packaging ideas that were so successful in aircraft to cars?

Concurrently, Elmer Paul of the Indiana State Police was gathering data from car injury crashes. His interests coincided with DeHaven's. Their collaboration led to the creation of the Automotive Crash Injury Research (ACIR) program at Cornell University Medical College. The researchers in this group were devoted to understanding the so-called **second crash** in a car accident.

Recall the crash that DeHaven witnessed in 1935. The driver lost control and the car skidded on a wet road, flipped over, and landed in a ditch. This is called the **first crash**. The cause of the first crash was the driver losing control on a wet road. However, the driver's severe injuries were caused by the force that threw him against the dashboard, where his face was ripped by the sharp, steel windshield wiper knob. The driver, or any occupant, tossed around inside the car is called the second crash.

In the mid-1950s, U.S. officials reported some 40,000 deaths and more than 1 million injuries from motor vehicle accidents. Nationwide, thousands of professionals, civil servants, and ordinary people focused on finding ways to reduce car accidents. These coalitions advocated mainly for the three E's: education, enforcement, and engineering. They wanted more people to receive driver education, increase policing to remove unlawful drivers, and better engineered roads. Essentially, the three E's promoted responsible driving and traffic management to prevent the first crash.

However, the ACIR group focused on reducing the harmful effects of the second crash. Fulfilling DeHaven's vision, they collected data to understand the specific **causes** of injuries in car accidents, as well as the **frequency** and **severity** of these injuries.

DATA AS EVIDENCE

The ACIR researchers adopted a medical approach to car crashes. In their view, these crashes were an epidemic. Ideally, if all car accidents could be eliminated, then the disease would be wiped out. However, a more realistic approach was to assume that a certain number of accidents would always take place. ACIR advocated for preventing injury when an accident did occur. They likened it to vaccines that protect people from the effects of exposure to infectious diseases.

ACIR's mission was to understand the relationship among the host (occupant) who had the disease (injury), the agent (component in the car) that transmitted it, and the environment (car) in which the host and agent existed. They believed that reducing injuries and deaths by improving components in cars was akin to bringing the disease under control long before the nation, as a whole, could do away with all car crashes.

To carry out their mission, the ACIR team set up the first nationwide and detailed data collection program on car crashes. They obtained voluntary cooperation from agencies within 21 states (Arizona, California, Colorado, Connecticut, Georgia, Illinois, Indiana, Maryland, Michigan, Minnesota, New Mexico, New York, North Carolina, Ohio, Oregon, Pennsylvania, South Carolina, Texas, Vermont, Virginia, and Wisconsin) and four cities (Cincinnati, Ohio, El Paso, Texas, Minneapolis, Minn., and Worcester, Mass.). The roadways were their laboratories. State traffic-control authorities, medical organizations, hospital personnel, practicing physicians, and highway patrol officers acted as laboratory assistants.

These data collectors recorded many key pieces of information about a crash, including answers to questions such as:

- How many vehicles were involved in the crash?
- Did the vehicle rollover?
- What was the estimated speed at impact?
- Where was the most damage on the vehicle?

Also, details about the vehicle included the make, model, year, and body style (such as sedan or convertible). Information about the occupants included where they sat, whether or not they were ejected, their sex, age, weight, and height. Police worked with doctors to record the cause and severity of occupant injuries. They recorded road and weather conditions. In total, more than 100 detailed elements of the crashes were collected. Photographs, diagrams, and narratives helped to verify the quality of the data. The information went into the ACIR database. By the mid-1950s, the ACIR had data on thousands of injury-producing crashes. Researchers could now report on what was harming and killing car occupants.

A MYTH-BUSTING STUDY

One of the first major studies by the ACIR program concerned occupant ejection in car crashes. In the 1950s, there was a common belief that occupants were more likely to survive a crash if they were "thrown clear" from the wreckage. The researchers wanted to see if this belief was true.

The group created a series of questions. First, they asked the safety question: "Were more occupants dying when ejected than when not ejected?" If they found that occupants were more likely to die when ejected, then an additional research question needed to be answered: "If ejected occupants had remained in the car, then how many lives would have been saved?"

From their database, they gathered relevant data on over 6,700 occupants. The researchers matched occupants by crash severity, seating position, and environment (urban versus rural roadways). They compared the injury patterns of those who were ejected and not ejected.

The data showed that the probability of dying was much higher if the occupant was ejected, at 12.1%, compared to if the occupant was not ejected, at 2.5%. In other words, it was better not to be ejected (Figure 2.2).

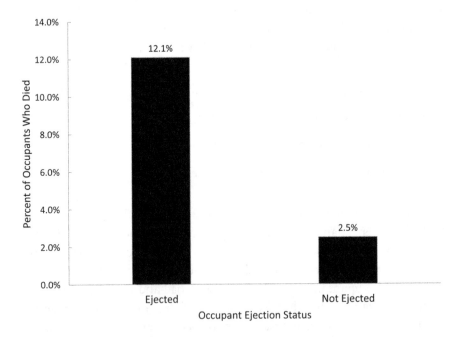

FIGURE 2.2 Results from ACIR study on occupant ejection. The data from the publication showed that the probability of dying was higher if the occupant was ejected.

However, the team could not confirm that the occupant's life would have been saved if s/he had remained in the car. For instance, if the car's roof severely crushed into the occupant compartment in a rollover crash, the non-ejected occupant may still have suffered from fatal injuries. Nonetheless, remaining in the vehicle proved to be the better option in general. The researchers concluded that over 50% of the fatally injured, ejected occupants would probably have survived if they were secured in their vehicles. That translated into about 5,500 lives annually.

The ACIR group concluded that ejection was a safety problem. Plus, they went one step further. They performed a detailed engineering analysis looking for the reasons for the ejections. Based on their findings, the group made three recommendations to the auto industry:

1. Improve door latches on future car models to prevent them from opening;

2. Promote the use of seat belts to keep occupants from being "thrown clear;" and

3. Find a simple and effective device for keeping doors closed on the 50 million, pre-1956 model vehicles currently on the road.

In summary, the ACIR program debunked the myth that it was better to be thrown clear from a vehicle in a crash. Their data showed that ejection increased the chances of dying. And the number one cause of the ejection was the door latch opening during a crash. To save lives, it was up to the car companies to strengthen door latches on future models.

A DATA STRATEGY DRIVES A NEW MOVEMENT

By the late 1950s, the Cornell researchers were receiving funding from the Commission on Accidental Trauma of the Armed Forces Epidemiological Board. That included funds from the Surgeon General, the Department of Army, and the National Institutes of Health. The Ford Motor Company and Chrysler Corporation also financially supported the group. The ACIR experts reported their safety findings in scholarly journals, at technical conferences, and in congressional hearings. The ejection study is one such published study.

However, the ejection study also exemplifies a bigger step forward in car safety. It illustrates a general strategy for using data to improve the safety of cars. Figure 2.3 highlights the approach. It starts with a suspected safety

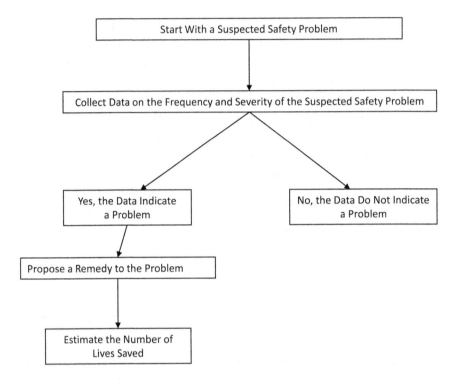

FIGURE 2.3   A strategy for using data in car safety investigations. The steps outlined here reflect the general data procedure used in most car safety debates.

problem, such as the issue of being ejected in a car accident. The next step is to address the questions originally posed by DeHaven: How often does this problem happen and how severe are the consequences? If the findings are significant, as the ACIR group found in their ejection study, then proceed to make some recommendations to fix the problem. Finally, use these results to estimate the impact, i.e., the number of lives saved, if the problem were fixed.

NOTE FROM THE AUTHOR

The enduring impact of this new approach to car safety cannot be overstated. DeHaven started with observing crashes in airplanes and went on to propel forward a new movement in car safety. Collecting data on car crashes became central to understanding how to make cars safer. And the data emboldened consumer advocates to press the car industry for radical change.

These consumer-orientated campaigns were often at odds with the young car industry, which was seeking to capture the consumer market and make money. Volumes have been written about the early years of this emerging technology. Surely, many private debates took place in corporate backrooms about the direction of the industry. What cars would sell. What the public wanted. Would safer cars sell? My purpose is not to discount or simplify those business challenges.

Rather, my main purpose is to illustrate the use of data in public safety debates. The work of the ACIR team serves as a perfect first example. If their research discovered a car feature caused injury, such as weak door latches, then the industry was expected to fix it. However, if the carmakers did not respond, then ACIR called on the federal government to act. As I briefly describe in Chapter 3, the issues championed by this group in the 1950s and 1960s laid the foundation for many of the car safety features that protect us today.

NOTE

1. Editorial (1937). The biologist looks at the motor car, *Scientific American* (April).

CHAPTER 3

Crash Data Make
a Difference

After a half a century of automobile usage, a staff of only nine people began, with federal support, the first statistical reporting system on how interior car designs injure and kill motorists. The time for analyzing the design of automobiles had come, and the crucial distinction between the causes of accidents and causes of injury was known with unmistakable clarity.

—RALPH NADER (1965)[1]

You approach your car and unlock it. You open the door, slide into the seat, and close the door, secure and comfortable. A buzzer issues a warning, and you fasten your seatbelt. You look out the windshield, turn on the windshield wipers or the defrost system, and adjust the rearview mirrors. Finally, you lay your head back on the headrest feeling like you are ready to go. You have just experienced the results of over 60 years of car safety research.

A key and controversial part of that research has always been data from car crashes.

DATA AS PROOF

In 1956, Dr. John O. Moore of Cornell's ACIR testified before a congressional committee on traffic safety. He was there to educate the lawmakers on data usage. He detailed the science behind the cause-and-effect of injuries in car accidents. His graphs were descriptive and convincing. He

DOI: 10.1201/9781003035343-3

FIGURE 3.1   An illustration of forces on the human body. Moore used these cartoons to describe the physics behind injuries.

identified the most dangerous features in cars. He told them how changes in car design could save lives and reduce injuries.

Moore's testimony began with a physics lesson. His illustrations are re-created in Figure 3.1. The first cartoon in Figure 3.1 illustrates using a large, flat rock on top of a man the concept of "peak load and duration of

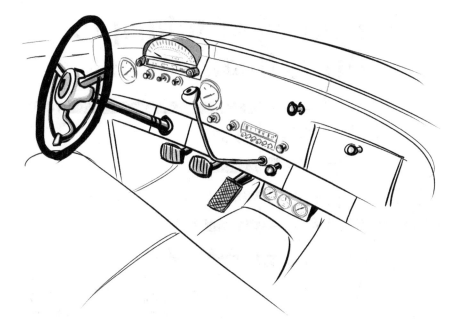

FIGURE 3.2   Illustration of a dashboard and steering wheel of a car, circa 1950s.

force." As Moore explained, the man can withstand the rock as long as its flat side lays against his chest. But what would happen if a sharp corner of the rock and its weight were concentrated on a small section of the man's body (Figure 3.1 "distribution of force"), and it was applied suddenly at a high rate of speed (Figure 3.1 "rate of onset")? These last two cartoons represent what would happen in a crash if the seating compartment in the vehicle was not well designed for safety.

Moore explained that sharp knobs and levers on dashboards localized, rather than distributed, the forces in a crash. These stylized elements were built-in safety hazards. As Figure 3.2 illustrates, car dashboards in the mid-1950s resembled early airplane cockpits. Moore's team had learned how to redesign cockpits for the U.S. Air Force with safety in mind. He made the case to the lawmakers that car occupants needed to be similarly packaged and protected to reduce injuries and deaths.

Moore tempered his message to the traffic safety committee by explaining that approximately 76% of occupants walked away from car crashes with no injuries or just minor ones. Much progress had been made in building safer cars since the early 1900s.

Frequency of Injury to Gross Body Areas

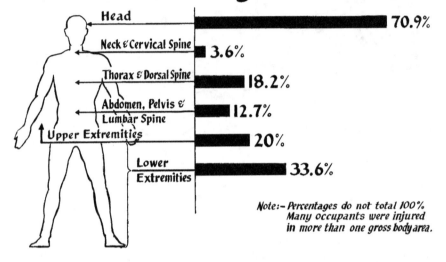

Head — 70.9%

Neck & Cervical Spine — 3.6%

Thorax & Dorsal Spine — 18.2%

Abdomen, Pelvis & Lumbar Spine — 12.7%

Upper Extremities — 20%

Lower Extremities — 33.6%

Note:– Percentages do not total 100%. Many occupants were injured in more than one gross body area.

FIGURE 3.3 Frequency of injury to gross body areas. Moore used this info-graphic to convince lawmakers to make cars safer.

However, the remaining 24% of occupants represented a very large number. Annually, up to 1.5 million people were injured and at least 30,000 people killed. The ACIR team had gathered data from thousands of car crashes. They knew from the data what features were injuring and killing occupants. In addition, they knew how to fix the problem.

When he got down to the specifics about injuries in crashes, he used the graphic shown in Figure 3.3. The occupants' heads had the highest frequency of injury. More than 70% or seven out of ten occupants received head injuries. The second most frequently injured body parts were the occupants' lower extremities, that is, legs and feet. About one in three occupants experienced these injuries.

Then, Moore spoke about the seriousness of these injuries, as shown in Figure 3.4. Head injuries were dangerous about 4% of the time and fatal an additional 5%. In other words, nearly one in ten head injuries were very serious. On the other hand, lower extremity injuries to the legs and feet were seldom dangerous or fatal (less than 1% of the time).

Seriousness of Injury Among 6 Body Areas

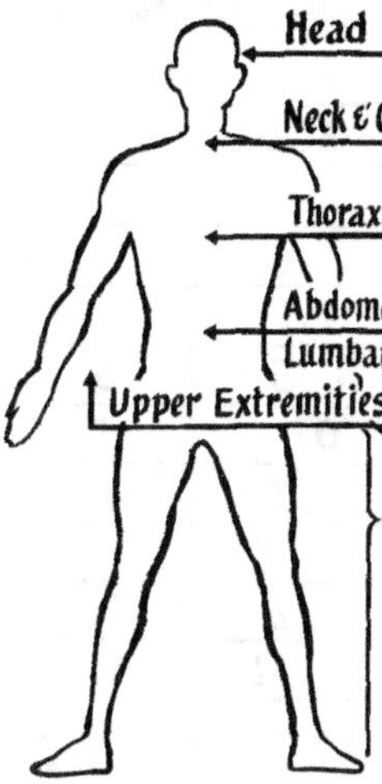

	Seriousness of Injury				
	Minor	Non-Dangerous	Dangerous	Fatal	Degree Not Reported
Head	56.2%	18.6%	3.7%	5.0%	16.6%
Neck & Cervical Spine	44.0	17.6	5.6	13.4	19.4
Thorax & Dorsal Spine	50.7	20.1	6.4	4.9	17.8
Abdomen, Pelvis & Lumbar Spine	46.0	18.9	12.9	3.5	18.7
Upper Extremities	71.7	20.2	.5	.1	7.5
Lower Extremities	75.4	17.1	.3	.1	7.0

FIGURE 3.4 Seriousness of injury among six body areas.

Moore concluded his testimony with a powerful list, as represented in Figure 3.5. The list was more than mere information – it was a call to action. The list contained the things that needed to be fixed in cars, ranked by "order of importance."

The top priority was to change the steering wheel. It needed to collapse or "give" in a crash, rather than spear the driver through the chest. The second priority was to prevent ejection by strengthening door latches. Thirdly, the instrument panel should become softer and rounder. And fourth, something had to be done about the windshield. The glass was cutting and dismembering drivers and passengers. The list continued, all the way down to making rearview mirrors less hazardous.

The level of detail in Moore's testimony was the hallmark of the ACIR's work. The researchers were fulfilling DeHaven's vision for improving car safety. They were providing the undisputable evidence needed to effect change. They had created a movement and now it was gaining ground.

Major Causes of Injury

	% of Occupants Injured to:			
	Any Degree	Moderate-Fatal Degree	Dangerous-Fatal Degree	Order of Importance*
Steering Assembly	29.4	8.4	2.5	1
Ejection	14.6	6.9	3.2	2
Instrument Panel	20.6	4.2	.7	3
Windshield	16.9	4.6	.6	4
Backrest of Front Seat (Top Portion)	11.0	2.4	1.1	5
Door Structures	7.7	2.4	.5	6
Backrest of Front Seat (Lower Portion)	15.1	2.5	0	7
Front Corner Post	2.0	1.2	.7	8
Flying Glass	3.0	.5	.02	9
Top Structures	1.2	.6	.2	10
Rear View Mirror	2.2	.6	.02	11

*Based on (a) the number of occupants actually exposed to the injury-hazard of the object, (b) the frequency of injury caused by the object, (c) the degree of injury caused by the object.

FIGURE 3.5 Major causes of injury. Possibly the most important table presented to Congress during the hearing, this gave the government and industry a list of priorities for re-designing cars to save lives.

OTHERS WEIGH IN

The ACIR's team and its statistical findings were cited over 100 times during the 1956 traffic safety hearings. For example, A.L. Haynes, executive engineer at Ford Motor Company, described the usefulness of the ACIR's data.

> Regardless of our success in designing accident prevention features into our product, the imperfect human element continues to precipitate accident situations. Therefore, our efforts also must be directed toward minimizing and preventing injuries suffered in these accidents.

> Statistics provided by Cornell University Medical College [ACIR] indicate the injury-producing areas of highest frequency. Experiments have shown that there is a range of forces at the

moment of impact which the human body probably can tolerate. Our problem then becomes that of providing methods of protecting car occupants involved in accidents that are not avoided.

Haynes also cited the experimental work of Colonel John Stapp, M.D., Ph.D., chief of Aero Medical Field Laboratory, Holloman Air Development Center. When Stapp spoke before the committee, he already had a reputation as the "fastest man on earth." As an Air Force officer, flight surgeon, physician, and biophysicist, Stapp was renowned for his novel studies of the effect of acceleration and deceleration on the human body. He based his findings on personal experiments on a test track. Stapp was one of the military volunteers who strapped themselves into a cart on a high-speed track to estimate the boundaries of human tolerance. They wanted to know what the strength requirements of restraint systems in aircraft were when they traveled at high speeds and stopped suddenly. Stapp reportedly survived one test run with forces up to 38 g.

When the committee asked Stapp to comment on speed as a factor in injuries, he gave the following response:

> People keep saying that speed kills. It isn't speed. It's the change of speed that can kill you. Going from 60 miles an hour to a stop in about 3 feet [is] nothing particularly arduous if you are properly fastened in, but it can be highly fatal if you aren't.

Testimony by Dr. Frank H. Mayfield, founder of the Mayfield Clinic and Spine Institute, suggested that the medical profession appreciated the data produced by the Cornell team and Colonel Stapp.

> In February, the American College of Surgeons passed a resolution [stating that] improvement in automobile design could materially reduce crash casualties and suggested ... that the data ... available from Cornell and other research sources indicated that improvement of the steering wheel, interior padding, restraining devices, door locks, and perhaps a seat restraint were avenues which could be approached.

An article titled "Big Three Fight Over How Safe to Make Your Car" was submitted to the traffic safety committee. The author of the article used the dispute about seat belts to highlight different views within the car industry. First, there were the fence-sitters. They held a freedom-of-choice point of view, "Wear the belts if you want to, but we're not sure they'll do much good." Second, were the advocates, such as Ford, Chrysler, engineers, doctors, and researchers who thought that well-designed seat belts would reduce death and injuries in most crashes, though not the most severe ones. Finally, there was General Motors, the hold out. According to the article, General Motors refused to provide safety belts and tried to block competitors' plans to offer them.

Several detailed quotes from industry leaders, including this one from Mr. James C. Zeder, Chrysler's engineering vice president, captured these attitudes:

> We benefited a great deal from the studies of such organizations as Cornell University, the Indiana State police, and the University of California at Los Angeles. But it may be years before we have really conclusive answers to the degree of added protection seat belts do afford. However, the findings of these and other reputable safety groups, together with our own laboratory and proving grounds test data, convince us that we now should make seat belts available to motorists who desire them.

The article's quotation from Mr. Fletcher Platt, Ford's traffic safety manager, also indicated that he accepted the findings of the Cornell group and Colonel Stapp, though with some reservations. He explained that Ford's position was that "any force that acts to restrain a person in a crash would tend to reduce the severity of injuries." Interestingly, he added another reason to wear a seat belt, "They help a driver keep control of his car when he's driving on a rough road or hits a boulder or pothole." Platt said the belts "are only a partial answer, but if you like to gamble, at least you'll have the odds with you."

The quotes from Howard Gandelot, General Motors' safety engineer, indicated skepticism: "General Motors hasn't said they're no good. We're just waiting to find out if they are any good. Nobody knows."

When asked about the ACIR's findings about ejection, injury, and door openings, again Gandelot appeared unconvinced:

> I don't know what to believe. You take all this talk about door openings. No one knows exactly what happens in an accident. A lot of people probably figure that the best thing is to get out of the car as

fast as they can. They reach over and open the door and fall out. Then the Cornell people tell us it's a door opening and they are thrown out.

More generally, the author of the article probed Gandelot about the broader issue of the sources of injuries in crashes:

A lot of people are hurt in bathtubs, aren't they? Do you hear anybody demanding that they take the bathtubs out of homes? A lot of people fall on hardwood floors and hurt themselves, don't they? Should we take the hardwood floors out of homes?

Gandelot cited automotive improvements, together with other advances, that had cut the highway death rate in half since 1938. When pushed further concerning ways to prevent the 30,000–40,000 deaths that were happening, Gandelot described GM's approach as "an honest difference of opinion." Then he added, "You can't blame us for that, can you?"

Similar congressional hearings continued for more than a decade. Safety advocates and the public voiced their concerns about the deaths, injuries, and destruction on our roadways. Then, a monumental change occurred.

THE IMPACT

In March 1966, President Lyndon Johnson proposed the creation of the Department of Transportation (DOT). In a speech before Congress, Johnson cited the costs and progress of our modern transportation system. He acknowledged the challenge of preventing deaths and injuries on the nation's roadways. He also echoed the indignation of the now famous congressional researcher and author, Ralph Nader. Remarkably, Johnson pointed directly to the important role of data, "We must acquire the reliable information we need for intelligent decisions."

Johnson signed into law the National Traffic and Motor Vehicle Safety Act of 1966, creating the DOT. In that same year, the Highway Safety Act tasked a new bureau within the DOT with reducing injuries, fatalities, and economic loss from crashes on our nation's roadways. The bureau would become the National Highway Traffic Safety Administration (NHTSA).

Dr. William Haddon, a physician-epidemiologist and engineer, became the bureau's first leader. He came to the job with extensive experience as a member of the New York Department of Health. Under his stewardship, the state had required all cars to meet a set of safety standards.

Haddon hit the ground running when he arrived in Washington in 1966. He led the charge to create the first Federal Motor Vehicle Safety Standards (FMVSS). The standards required carmakers to build in certain minimum performance criteria for protecting occupants. Certain safety features would no longer be optional for the auto industry.

By 1967, several of these standards – long in the making – were finalized and became effective in 1968:

FMVSS 201: Occupant Protection in Interior Impact. This standard required makers to use padding and other types of protection for occupants in the event of a crash.

FMVSS 203: Impact Protection for the Driver From the Steering Control System. This standard mandated the design of a steering control system to protect drivers from chest, neck, and facial injuries in a crash.

FMVSS 206: Door Locks and Door Retention Components. To reduce the chances of occupants being ejected from a vehicle in the event of a crash, this standard required stronger latches, hinges, and other door equipment.

FMVSS 208A-E: Lap Belts.

Federal regulations define requirements intended to protect the public against unreasonable risk of crashes occurring due to design, construction, or operating performance of motor vehicles. In addition, they are also intended to protect the public against unreasonable risk of death or injury in the event of a crash.

Haddon's tenure was short. In 1969, he left the federal government and soon became the president of the Insurance Institute for Highway Safety (IIHS), an industry supported group. As you will see in later chapters, this organization has greatly advanced car safety with a heavy reliance on data. Haddon's legacy from these roles places him on par with DeHaven as a pioneer in the scientific, data-intensive approach to car safety.

NOTE FROM THE AUTHOR

There is a clear link between these early data analyses and the far-reaching federal safety standards in place for our cars today. And as car safety has evolved, so has the use of data.

When Moore made his presentation before Congress, the ACIR essentially held all the cards. They had detailed analyses about thousands of crashes. There was no other national, public dataset to counter their analyses or findings.

Today, it's a different story. The federal government collects information on thousands of crashes annually and makes that data public. In addition, there are datasets of police accident reports from the states and datasets from hospital emergency rooms. Plus, modern vehicles have a so-called black box, which contains streams of data on a car's components, such as braking and steering.

With all these datasets available, every safety problem identified by NHTSA is likely to be challenged. Organizations, especially the carmakers, often choose a different dataset or type of analysis to reach an alternate conclusion. Herein lies the heart of the debates. I will delve into the use of data in these disputes in the following chapters, but first I need to acquaint you with the most important dataset, the Fatality Analysis Reporting System.

NOTE

1. Nader, Ralph. (1965). *Unsafe at any speed: The designed-in dangers of the American automobile.* Grossman Publishers, p. 133–134. The quote refers to the ACIR group.

CHAPTER 4

Measuring Progress
with Data

The purpose of crash data is to help decision-makers understand the
nature, causes, and injury outcomes of crashes. This information
provides context for the design of strategies and interventions that
will reduce crashes and their consequences.

—UMassSafe (2018)[1]

SANTA CLARITA, Calif.–U.S. actor Paul Walker, 40, who starred
in *The Fast and Furious* movie franchise, died in a single-vehicle
crash on November 30, 2013. The Los Angeles County Sheriff's
department said Walker was in the passenger seat when his friend,
Roger Rodas, lost control of the 2005 Porsche Carrera GT and
crashed the vehicle into a tree. The car exploded into flames, killing both. Speed may have been a factor.

In addition to widespread coverage in newspapers, broadcasts, and social
media, Paul Walker and Roger Rodas' fatal car crash appears in a federal
database. Nearly 32,900 other people died in motor crashes in 2013. They
are also in this database – nameless and digitized for analysis.

In 1975, the National Highway Traffic Safety Administration (NHTSA),
a branch within the federal Department of Transportation, introduced a
data collection system titled the Fatality Analysis Reporting System also
known as FARS. This database contains detailed information about all
fatal motor vehicle crashes on our nation's public roadways.

DOI: 10.1201/9781003035343-4

Experts consider the data in FARS the gold standard in car safety. The status is due to the fact that FARS is a census of fatal crashes and that the information comes from multiple sources. Within each state government, there is an individual or individuals responsible for following up on a fatal crash. It's their responsibility to submit a complete record to the central repository at the federal level. The information stems from various official sources:

- police accident reports

- state vehicle registration files

- state driver licensing files

- state highway department data

- vital statistics

- death certificates

- coroner/medical examiner reports

- emergency medical service reports

- and other state records, as needed

For example, there are more than 200 pieces of information about the Walker–Rodas crash in FARS.

Nearly everyone in the car safety community turns to FARS to measure safety progress. However, it's more complicated than looking at simple counts of fatalities. If the number of fatalities on our roadways decreases, can we call this progress? Can we declare victory that certain countermeasures, such as seat belts or drunk driving laws, work?

In this chapter, we focus on understanding the context of these numbers, what may lie behind the numbers, and how they might be interpreted. These data-intensive discussions provide an important backdrop for the debates in car safety described in later chapters of this book.

COUNTS AND RATES

At its most basic level, FARS provides annual **counts** of deaths on the roadways. Each year, NHTSA publishes summary graphs of these counts. Figure 4.1 is an example. It's a plot of the number of motor-vehicle-related fatalities over the past 45 years. The graph makes it easy to see the general

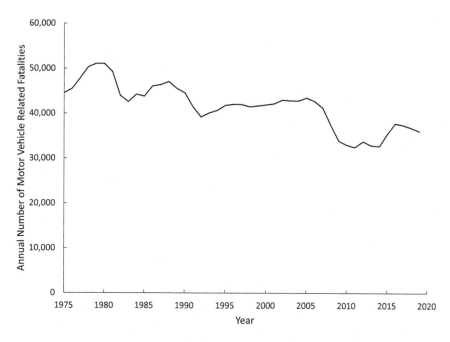

FIGURE 4.1 Annual number of motor vehicle-related fatalities, 1975–2019.

decreasing trend in the number of deaths, from more than 44,500 in 1975 to less than 36,100 in 2019. This represents nearly a 20% decrease since the mid-1970s.

Counting fatalities to measure the progress in overall safety is consistent with the idea introduced by DeHaven. He wanted to know how many people were injured and killed in car crashes. He wanted to reduce these numbers. The ACIR team at Cornell followed a similar approach when they reported on the number of accident victims. In their day, the data in car safety focused on the *numbers* of lives affected. The number of people who could be saved if vehicles were redesigned. The actual counts were meaningful to people in these early years.

However, as the number of vehicles on the road reached into the hundreds of millions, a new way of looking at the data occurred. Instead of counting the number of fatalities or injured occupants, some in the safety community started to use rates. The idea was to scale the data to take into account **opportunity** or **exposure**. After all, if there were more vehicles or licensed drivers on the road, then there would probably be more deaths. They felt it was more meaningful to use counts **relative** to some measure of exposure versus the counts alone.

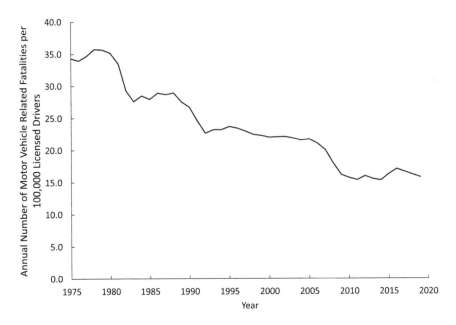

FIGURE 4.2   Licensed driver fatality rates, 1975–2019.

For example, consider the number of licensed drivers. All drivers are exposed or have the opportunity to be in a fatal crash. Therefore, it is logical to look at the fatality counts relative to the number of all licensed drivers. This is computed as:

$$Licensed \ Driver \ Fatality \ Rate = \frac{Number \ of \ Fatalities}{Number \ of \ Licensed \ Drivers}$$

Figure 4.2 shows this annual rate over the 45-year period. Similar to the fatality counts in Figure 4.1, this graph has a general downward trend. However, when it comes to interpreting this pattern, it gets more complicated. Rates, especially rates computed over long periods of time, are influenced by social, economic, and marketing trends.

Figure 4.3 is a plot of the female-to-male composition of licensed drivers over the past 45 years. Since 2005, the number of female drivers exceeds male drivers. This change is important because female drivers have better overall safety records than men. Specifically, men are **three times more likely** to be involved in a fatal crash than female drivers. This factor of three has been consistent for the past ten years. So, while the downward pattern seems to signal progress in car safety, the actual cause may be the change in the type of drivers. More female drivers lead to fewer fatalities and better safety rates.

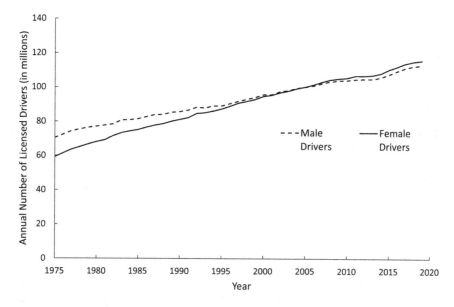

FIGURE 4.3   Number of licensed drivers in millions, female vs. male, 1975–2019.

If the goal is to reduce this rate further, then an extreme step would be to pass a law that licenses only female drivers. This would cut the number of fatalities by **about half**. In other words, the most effective car safety policy might be to take all men off the road. This tongue-in-cheek idea illustrates the folly of using a rate to claim progress in car safety.

Now consider these fatality rates:

$$Registered\ Vehicle\ Fatality\ Rate = \frac{Number\ of\ Fatalities}{Number\ of\ Registered\ Vehicles}$$

and

$$Miles\ Driven\ Fatality\ Rate = \frac{Number\ of\ Fatalities}{Number\ of\ Miles\ Driven}$$

The graphical display of these rates in Figures 4.4 and 4.5 shows a decreasing trend, similar to Figure 4.2. But again, use caution in interpreting the pattern. First of all, we own more vehicles now. According to federal statistics, private vehicle ownership has increased from 1.6 vehicles per household in 1977 to 1.9 in 2017. Plus, the category of light trucks – which includes pickups, sport utility vehicles, and minivans – appears to be the

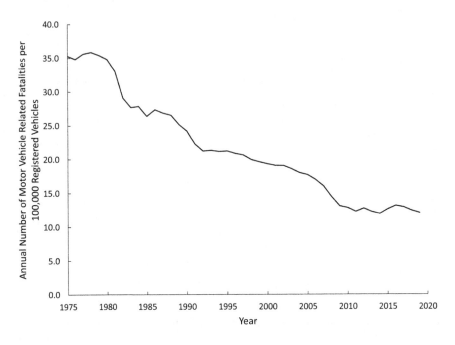

FIGURE 4.4   Registered vehicle fatality rates, 1975–2019.

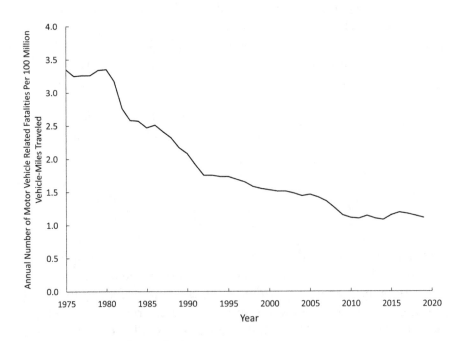

FIGURE 4.5   Miles driven fatality rates, 1975–2019.

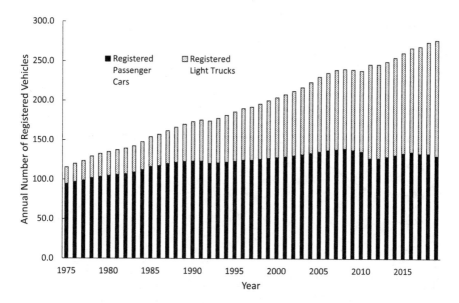

FIGURE 4.6 Annual number of registered vehicles, passenger cars, and light trucks, 1975–2019. Cars outnumbered light trucks four to one in 1975. There were more light trucks on the road in 2019 than cars.

main source of that growth, as shown in Figure 4.6. In 1975, passenger cars outnumbered light trucks by more than 4 to 1. In 2019, there were more of these heavier vehicles on the road than passenger cars. While heavier vehicles are typically safer vehicles for the occupants, they pose bigger threats to people in smaller, lighter cars. Consequently, from a car safety standpoint, light trucks are both a positive and a negative.

What about the rate using miles traveled in Figure 4.5? Many analysts prefer this measure because it takes into account the actual usage of the vehicles. Again, be cautious in interpreting the rate trend. For example, over this period the U.S. population moved out of the countryside and into the cities. In 1960, about 37% of the population lived in rural areas. By 2010, that percentage had decreased to 20%. Miles driven in urban areas are slower than rural miles, primarily due to congestion. As a consequence, there are fewer fatal crashes per vehicle-miles traveled on these urban roads than rural ones. Now, add in the fact that between the early 1980s and the present, female drivers increased their average annual miles from about 6,400 to nearly 10,000, as shown in Figure 4.7. Taken all together, the progress in safety as measured by these rates may be due to who is driving (women) and where they are driving (urban areas), rather than real improvements in automotive safety.

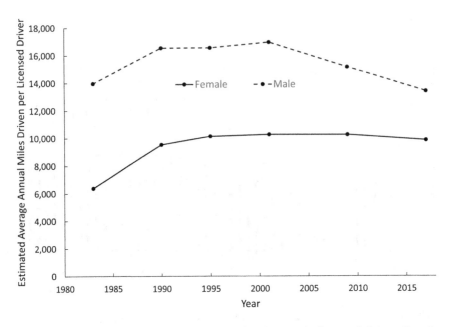

FIGURE 4.7   Estimated average annual miles driven per licensed driver, female vs. male, years: 1983, 1990, 1995, 2001, 2009, 2017.

By now, you have probably figured out that we need to be very careful when measuring safety using rates. The problem is that the divisor, the number used to normalize or scale the counts, may have its own pattern. And when we mix the pattern of the fatality counts with the pattern of the normalizing factor, assigning credit for the downward trend in these graphs is controversial.

However, returning to the counts shown in Figure 4.1, there is the possibility of crediting some national campaigns to the decreasing patterns.

SEAT BELT AND ALCOHOL-USAGE POLICIES AND LAWS

Since the early days of DeHaven and the Cornell team, experts have recognized **seat belts** as an important way to reduce injury and save lives in crashes. However, there are two requirements that must exist in order for seat belts to be effective. First, the belts have to be in the vehicle. Second, people have to wear them. And these two factors did not happen overnight.

Some manufacturers installed lap belts in cars in the 1950s and 1960s. However, early sales teams thought the presence of seat belts would hurt sales. Salespersons often tried to hide these devices from the consumer. Belts lying across the seats might remind buyers of the safety risk of driving their new car.

Nonetheless, a federal standard in 1967 required all carmakers to install these safety devices for the driver and right front passenger. The public no longer had a choice when buying a new car. As technology advanced, the government amended this standard in 1973 to require integral 3-point belts and in 1974 to require a 4- to 8-second audible reminder to buckle up. The government delayed the requirement of lap belts in light trucks until 1970 and 3-point belts until 1975.

Initially, the public resisted wearing seat belts. The common view was that the government could not force people to buckle up. It violated basic freedoms. Furthermore, such a law would likely be unenforceable.

While the United States debated this issue, new data arrived from Down Under. Australia's southern state of Victoria passed a law in 1970 that required people to wear seat belts. By 1974, motor vehicle related deaths there had dropped by more than 33%. In other words, Victoria credited seat belts with saving one in three lives in car crashes.

It took ten years for any state in the United States to catch up. In 1984, New York passed the first mandatory seat belt law. Other states followed.

So, did people actually obey these laws? Apparently, they did, but it took time. Figure 4.8 shows the estimated seat belt use in the United States for

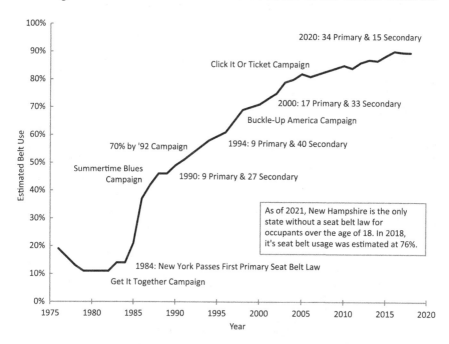

FIGURE 4.8   Estimated seat belt usage, 1976–2018, and the number of states with primary and secondary laws.

the period 1976 through 2018. National awareness campaigns and state primary and secondary seat belt laws began to take hold, as Figure 4.8 also shows. A primary seat belt law allows a police officer to ticket a driver or occupant for failure to wear a seat belt, without any other violation. In contrast, a secondary seat belt law requires that there be another traffic infraction as the reason for stopping the vehicle. Then, if the occupants are not wearing their seat belts, a citation can be issued for that as well. These laws along with federal buckle-up campaigns worked together to educate the public about the benefits of seat belts. As a result, lives were saved. According to federal research published in 2015, nearly 330,000 lives were saved between 1960 and 2012 by this single safety device.

Nearly every one of us knows someone who has been touched by tragedy involving a drunk driver. The sad fact is that drunk drivers are a big problem, but there has been some improvement over the years. In 1982, more than 21,200 people died in crashes involving an alcohol-impaired driver, nearly one in two recorded fatalities. (Note the alcohol-impaired driver may or may not have caused the fatality, but was counted among the drivers involved in the crash.) Since 1982, Figure 4.9 shows the number of people killed in these

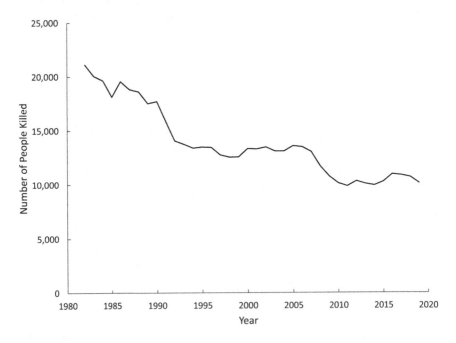

FIGURE 4.9   Number of people killed in crashes involving an alcohol-impaired driver, 1982–2019.

types of crashes has decreased. That is the good news, but alcohol is still a major safety problem on our roads.

In 2019, over 10,000 people died in crashes involving impaired drivers, nearly one in three fatalities. That is, about 27 deaths a day involve a drunk driver – about one every 53 minutes. Drivers are labeled drunk or alcohol impaired if their blood alcohol concentrations (BAC) are .08 grams per deciliter (g/dL) or higher. By 2005, every state had a law that made it illegal to operate a motor vehicle at or above .08 g/dL. On December 30, 2018, Utah enacted a stricter threshold of .05 g/dL or higher. Furthermore, operating a commercial vehicle with a BAC of .04 g/dL or above violates federal regulations and may result in criminal charges. It is reasonable, looking at Figure 4.9, to say that these laws have had a positive effect.

In addition to outlawing alcohol-impaired driving, there was a national movement to increase the legal drinking age in the early 1980s. The National Minimum Drinking Age Act of 1984 mandated that all states raise their minimum purchase and public possession of alcohol age to 21 as a condition for receiving state highway funding. By 1988, the minimum drinking age was uniform across the country.

At the same time, civic organizations such as Mothers Against Drunk Driving and Students Against Destructive Decisions (founded as Students Against Drunk Driving) were promoting sobriety behind the wheel. By 2008, all states and the District of Columbia had zero-tolerance laws making it illegal for drivers under 21 to operate a motor vehicle with a BAC of .02 g/dL or higher.

Together, such social campaigns appear to have had some effect on the youngest drivers' involvement in fatal crashes, as shown in Figure 4.10. The rate in this graph is computed in the following way:

Driver Involvement Fatal Crashes Rate

$$= \frac{Number\ of\ Drivers\ in\ FARS\ in\ Age\ Group}{Number\ of\ Licensed\ Drivers\ in\ Age\ Group}$$

Figure 4.10 reveals some interesting involvement patterns when you compare the two factors of sex and age:

- The youngest drivers, especially the men, have the highest involvement rates, probably a contributing factor to their higher insurance rates.

- Women's involvement rates are always less than that of men's by 15–30 percentage points.

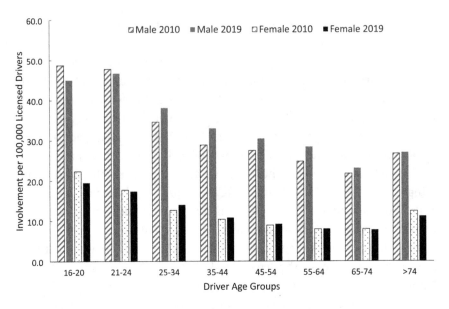

FIGURE 4.10 Driver involvement fatal crash rates by age groups, female vs. male, 2010 vs. 2019.

- Among middle-age drivers, men show an increase in involvement between 2010 and 2019, but women do not.

- Drivers more than the age of 74 have a relatively high involvement rate, particularly given that they tend to drive less.

These age-specific rates highlight the need to have age-specific countermeasures to improve the safety of our roads. For example, states have graduated teen licensing that restricts the time of day when a young driver can be on the road or restricts the number of young occupants in the car to reduce harm to these young drivers and others on the road. At the other end of the age range, the movement to require older drivers to renew their licenses more frequently reflects concern about this group's high rate of involvement in fatal crashes.

In summary, seat belt usage and alcohol-curbing laws have helped to reduce the number of motor vehicle-related fatalities. However, when we look at the data by age and sex, across the most recent nine years, these factors do not seem to have had a big effect. As you will see in later chapters, there may be other vehicle-related factors thwarting real progress in car safety.

NOTE FROM THE AUTHOR

I intended to overload you with data in this chapter. I wanted you to feel the complexity of interpreting the data patterns of fatalities on our roadways. The contributing factors discussed here are only a few of the many trends that could help explain the *why* behind the data:

- federal regulations: safety standards, like collapsing steering wheels and installed seat belts;

- state laws: mandating seat belt usage and lowering thresholds for alcohol-impairment violations;

- individual choices: what vehicle to buy, whether or not to wear a seat belt or to drive impaired;

- changes in the demographics of drivers: who and where people drive.

Needless to say, many researchers have dug much deeper into these factors. They mined various data sets to suggest more complex explanations behind these trends. Furthermore, carmakers have introduced new vehicles models with updated safety features, such as improved air bags and braking systems. Taking all these factors into account and delving into that research goes beyond my purpose in this chapter.

Rather, my goal is to give you some awareness of what might be behind the trends over the past 40+ years. It's important because, as you will see in the debates in the following chapters, industry leaders sometimes use factors such as these to argue against making changes to their vehicles. For instance, Jeep blamed excessive speed for the fire that enflamed Tommy's Grand Cherokee in the story in Chapter 1. Driver behavior is often cited as the problem.

In my opinion, poor driver behavior should not be an industry loophole; it is a design parameter. Vehicles should be designed knowing that they will be driven by flawed individuals. And, that we are all sharing the road.

What can actually be done to protect occupants in the event of a crash? The data tell us what flaws need fixing on cars to save lives and, moreover, where new technology could help prevent crashes, altogether.

NOTE

1. UMassSafe. (2018). *Why crash data is important.* Massachusetts Law Enforcement Crash Report E-Manual. https://masscrashreportmanual.com/data-importance/why-crash-data-is-important/ Accessed December 8, 2021.

CHAPTER 5

The Roof Crush Resistance Debate

All of us has known relatives or friends who were injured or killed in car crashes. Likewise, the statistical frequency of such events makes it possible for analysts to link effective regulation to changes in the frequency of such events.

—JOHN D. GRAHAM (1989)[1]

On August 23, 2005, National Highway Traffic Safety Administration (NHTSA) published its findings on how people were dying in rollover crashes: Roofs were crushing the occupants to death. The NHTSA administrators proposed changes to the federal safety standard for roof strength. Carmakers should strengthen the roofs of cars. Stronger roofs could withstand the roof-to-ground contact in rollover crashes and reduce occupant deaths and serious injuries.

When our federal government proposes such a change, it opens a public docket to solicit feedback. This docket generated hundreds of submissions. For Don Rank, who submitted the following handwritten letter to the docket, the matter was very personal.

DOI: 10.1201/9781003035343-5

SEPT. 6, 2005
LEWISTON, MI

Dear Sir:

I am writing to comment on a proposed new regulation that would require automakers to build stronger roofs on cars to protect people in rollover accidents.

I strongly support the new rules.

My younger brother, Clarence, was killed at age 70 in a rollover on I-40 near Winslow, AZ in August 2001. He was in his seatbelt & the Buick landed on the roof over his head & collapsed inward killing him instantly.

The auto industry is asserting that the seat belt protects a person in a rollover, but this allegation is patently false in cases where the car roof collapses. My brothers (sic) case is proof of the matter.

His wife was sleeping & in her seat belt & suffered only minor injuries because the car roof on her side did not cave in. So, the seat belt did protect her as long as the roof did not collapse inward.

The auto industry is aware of these facts & is trying to defuse the issue by making incongruous hypothesizes.

It is also my opinion that the rollover tests now being used are archaic. They need to be brought up to date. The recommended test is a dynamic one that rolls moving vehicles as in a real-life scenario. Not a static test, as now pushes over stationary cars. This is not using common sense with all the current testing equipment available.

I appreciate the opportunity to comment on these proposed new regulations. They are very important. After all, the industry defends itself by asserting their vehicles meet all regulations. This tactic has been very successful in the courtrooms.

Thank you.

SINCERELY,
Don Rank

The letter stands out in the docket because it gave the debate a personal perspective. Behind each dataset, hundreds of stories of personal tragedy could be told. However, in the usual course of arguing car safety issues, the focus is on the data and not the stories.

THE SAFETY PROBLEM AND A THEORY

Using nationwide statistics, NHTSA showed that rollover crashes were the most lethal. As Figure 5.1 illustrates, they had the highest rate of fatal and serious injuries. To reduce this human toll, roofs needed to be strengthened.

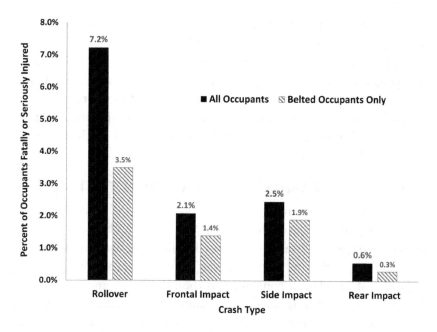

FIGURE 5.1   Fatally or seriously injured rates by crash type, 1997–2002. The high rate of rollover crashes served as justification for the proposal to strengthen roofs in cars.

NHTSA's idea behind stronger roofs is fairly simple, as illustrated in Figure 5.2a. Rollovers cause roof crush. The crush or intrusion reduces the occupant's compartment space, as illustrated in Figure 5.2b. With less space, occupants are more likely to be killed or injured. If roofs were strengthened, then they would be crush resistant. Hence, the Federal

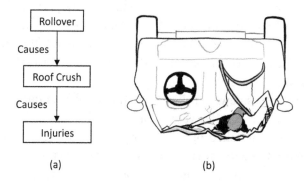

FIGURE 5.2   (a) Flowchart of NHTSA's general idea of the causal link between roof crush and occupant injuries in rollover crashes. (b) Illustration of how the restrained occupant is injured when the roof intrudes into the passenger compartment.

Vehicle Safety Standard (FMVSS 216) associated with this proposal is called the Roof Crush Resistance Standard.

Rank's letter was among the hundreds of documents NHTSA received from people, organizations, companies, and state governments. Some supported the proposed change and argued for even stronger roofs. Others argued against the proposal. Let's take a closer look at the important elements of a rollover crash, and the key feature of the roof crush standard at the center of the debate.

ROLLOVER CRASHES AND ROOF STRENGTH TESTING

Rollover crashes are complex events. Many factors influence the degree to which a rollover injures the occupants. Occupant characteristics and behavior, such as age (we become more fragile as we age) and sex (men's bodies tend to withstand impacts better than those of women), or being properly restrained, can contribute to the injury outcome. In addition, vehicle factors – such as the vehicle weight, size, structural design, size of the occupant compartment, padding, and airbags – can influence the degree of occupant injuries.

Last, but not least, is the nature of the crash. Researchers describe rollover crashes by the number of rolls, the number of roof-to-ground contacts, what initiated the rollover, the distance traveled, the speed of the roll, and what stopped the roll. The possible combinations of all these contributing factors to the occupants' injuries are almost endless. As will be discussed later in this chapter, these factors play an important role in the debate about data analysis in rollover crashes.

Since a rollover is a complex event, let's start with some basic definitions. Experts call a crash a rollover if, during any part of the crash, a vehicle rotates 90 degrees or more. This rotation can happen in two ways: The vehicle does a sideway roll or it does an end-to-end flip. The sideway roll is by far the most common (more than 90%), and for that reason, it is the focus of this discussion.

Experts describe sideways rollovers by the number of quarter turns, as illustrated in Figure 5.3. If a vehicle only rolled to one side, then it had **1 quarter turn**. If it rolled onto its roof and stops, then it had **2 quarter turns**. With **3 quarter turns**, the vehicle landed on the second side. If the vehicle rolled completely and landed on its wheels, then it experienced **4 quarter turns**. And, so forth.

Figure 5.3 also illustrates that when a vehicle only had a 1 quarter roll, there was no roof-to-ground contact. A vehicle must have had more

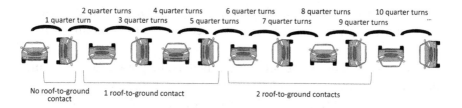

2 quarter turns   4 quarter turns   6 quarter turns   8 quarter turns   10 quarter turns
1 quarter turn   3 quarter turns   5 quarter turns   7 quarter turns   9 quarter turns   ...

No roof-to-ground   1 roof-to-ground contact   2 roof-to-ground contacts
contact

FIGURE 5.3   Illustration of sideways rollovers with quarter turns and roof-to-ground contact.

quarter turns to have the *potential* for roof-to-ground contact. The word *potential* is used because unless the vehicle landed on the roof, there was no absolute certainty that there was roof-to-ground contact. For example, a vehicle could go airborne during the roll, not make contact with the ground and experience no roof crush.

Roof-to-ground contact is at the heart of the Roof Crush Resistance Standard. To pass the standard, a car must show that its roof can withstand a simulated roof-to-ground contact and not crush into the occupant compartment.

Figure 5.4 illustrates that during the test, researchers place a large steel plate on the roof of the vehicle over the driver. They steadily increase the force of the plate until it is "equivalent to 1.5 times the unloaded weight of the vehicle." In order for a vehicle to pass or conform to this standard, its roof must offer sufficient resistance to the steel plate and not crush more than 5-inch inward under the force. While this test procedure has evolved

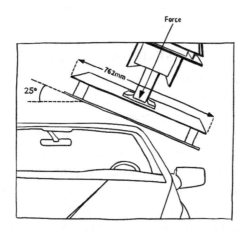

FIGURE 5.4   Illustration of test procedure for roof crush resistance standard.

since the first roof-crush standard in 1973 and contains many details, the single most important detail for this discussion is the amount of force applied.

In the description of the procedure, there is the phrase "a force equivalent to 1.5 times the unloaded weight of the vehicle." A quick example might help to clarify why engineers often measure force relative to the vehicle weight. Consider a crash in which the vehicle rolled onto its roof. All the weight of the vehicle is pushing into the roof. To prevent roof crush into the occupant compartment, the roof would need to withstand at least the weight of the vehicle. Analysts use the term **strength-to-weight ratio** (SWR) to capture this notion. You might think of the 1.5 amount as a safety margin, accounting for the dynamic forces in a rollover.

NHTSA proposed three options in changing the standard. The force applied by the steel plate in the procedure would increase to either 2.5, 3, or 3.5 times the unloaded vehicle weight. Also, to pass the test, there could be no intrusion into the head or neck of a seated dummy – intended to represent a front-seated occupant. In a nutshell, to conform to the new standard, vehicle roofs would need to be stronger.

In 2005, NHTSA used the data shown in Figure 5.1 to illustrate that a problem existed with rollover crashes. In addition, federal analysts supplied a statistical analysis to support the recommendation that stronger roofs would save lives. However, after two years of public debate, NHTSA published another statistical study and extended the feedback period on the docket, rather than issue a decision. The new study strongly polarized the debate between the parties opposing the new standard and those supporting it. A key point of contention was the data analysis.

NHTSA'S STATISTICAL ANALYSIS

NHTSA analyst Dr. Alexander Strashny authored the 2007 statistical study that served to extend the public docket. Rather than using the Fatality Analysis Reporting System datasets discussed in Chapter 4, Strashny used information from the National Automotive Sampling System–Crashworthiness Data System (NASS-CDS). This dataset consisted of detailed information of over 8,000 fatal and non-fatal rollover crashes. Key to this analysis were specifics about the rollover crash, including the number of quarter turns, the amount of intrusion over the occupant, the severity of occupant injuries, and the source of those injuries.

With these details about a crash, Strashny was able to comb through the NASS-CDS dataset and select only those rollover crash vehicles that were more likely to have benefited from stronger roofs. For instance, he excluded rollovers with only 1 quarter turn from the analysis because they did not have roof-to-ground contact. Also removed were vehicles in "arrested" rollovers. That is, vehicles stopped by a wall, guardrail, or tree. In these crashes, an occupant's injuries could be associated with these final impacts and not roof-to-ground contact. Furthermore, Strashny selected only single-vehicle crashes. This removed the complication of vehicle-to-vehicle contact.

Finally, and perhaps most importantly, Strashny focused on injuries caused by roof contact. More specifically, only head, neck, and face injuries identified as **caused** by contact with a roof component were included in his analysis.

The results of this new dataset and statistical analysis provided a clear link between roof intrusion and occupant safety. The conclusion was: **In rollover crashes, more roof intrusion led to increased risk of death or serious injuries.**

This general relationship is illustrated in Figure 5.5. More intrusion leads to higher chances of fatal or serious injury. In order to save lives and reduce injuries, roofs needed to be strengthened.

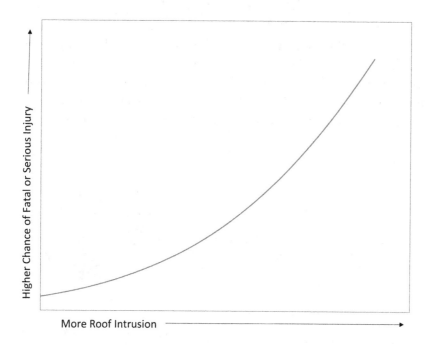

FIGURE 5.5   Illustration of results from NHTSA's 2007 statistical study. In rollover crashes, more roof intrusion increases the chances of fatal or serious injuries of an occupant's head, neck, or face.

While this latest study reinforced NHTSA's position, it also revealed some other important factors influencing occupant injury risks:

- Being a woman increased the chances of injury.

- Being older increased the chances.

- If the occupant was in a car, rather than a pickup, SUV, or minivan, then the chances of injury increased.

- More quarter turns in a crash increased the chances of injury.

Each factor, alone and grouped together with other factors, altered the chances of injury in rollover crashes. But overall, the conclusion remained the same – more roof intrusion led to a higher risk of injury.

There were few detractors of the 2007 study, itself. Rather, the loudest criticism came from industry representatives who disagreed with the NHTSA philosophy behind the analysis. As a result, they also disagreed with the conclusion.

AN ALTERNATE THEORY

Carmakers argued that NHTSA's theory about the link between roof crush and injuries was wrong. In their view, the rollover caused both the injuries and the roof crush, as illustrated in Figure 5.6a. During the roll, occupants "dove" into the roof, as shown in Figure 5.6b. The occupant would come into contact with the roof regardless of the roof strength or crush. In their view, strengthening the roof would not reduce deaths and injuries.

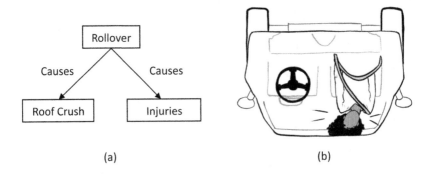

(a)   (b)

FIGURE 5.6   (a) Flowchart of industry's general idea of the causal link of injuries in rollover crashes. (b) Illustration of how the restrained occupant "dives" into the roof, even when there is no roof intrusion.

Industry representatives wrote letters that explained their position for the docket. They held that the severity of the rollover – such as the number of quarter turns, amount of crush, and distance the vehicle rolled – determined the injury, not the roof strength. NHTSA analysts made special mention of these perspectives in their Final Rule:

> Ford stated that, "The amount of roof deformation [crush] is only an indication of the severity of the impact between the roof and the ground."

> General Motors stated that, "Observations of injury occurrence at the end of a rollover collision reveal nothing regarding the relationship of roof deformation, roof strength, or roof strength-to-weight ratio injury causation."

> Nissan stated that deformation and injury severity are both independently associated with roof impact severity.

They also supported their theory with data from a set of industrial experiments carried out decades earlier.

In the mid-1980s, researchers and consultants at General Motors performed experiments focused on rollovers, roof strength, and injury causation. Commonly called the Malibu tests, these experiments used eight Malibu sedans from the model year 1983. Four of the sedans had a normal roof. The other four cars had roofs equipped with a roll cage structure (resembling a race car cage). Using a mechanical device to initiate a rollover event and laboratory recording devices, the engineers studied what happened to the belted dummies during the crash.

Based on the onboard camera video, the buttocks of the dummies "lifted off the seat cushion early in each test and never returned to the original depressed seat position until the test was completed." According to the engineers, impacts resulting from this diving-type movement caused "neck loads" or forces to the neck. This occurred when the head stopped and the torso's momentum compressed the neck. But they also found that, "Roof deformation never caused the dummy to be compressed between the roof and seat."

The researchers did find that dummies in the roll-cage vehicles had fewer, potentially injurious impacts and lower average neck load than their counterparts in the production vehicles. But this was attributed mainly to the difference in the amount of ground contact. According to these investigators, when the vehicles had similar ground impact, the roll-cage vehicles offered no better protection than the production vehicles. As a result, they held that stronger roofs made no difference in the injury outcome of the dummies.

The industry took these video-documented results and showcased them as evidence supporting their theory about how rollovers injured or killed occupants. The cause of injuries and death was crash severity, such as number of quarter turns, speed of the rollover, distance traveled, and the occupant's diving action, but not roof crush.

Not everyone agreed with this interpretation. For instance, critics took issue with how the researchers recorded the roof crush measurements and how different the neck of a dummy was from that of a human. Suffice it to say there were supporters and detractors of these experiments from the very beginning.

Given these two opposing theories about how occupants are injured in rollover crashes dating back decades, it was inevitable that statistical studies supporting both sides would become part of the debate to increase the roof strength test in the federal standard.

THE DATA DEBATE

Several non-NHTSA studies were submitted to the federal docket during the debate. Two distinct groups authored the most significant statistical studies. On the one hand, there were the longtime industry consultants Jaya Padmanaban and Edward Moffatt. Both served as expert witnesses for the automotive industry's legal disputes. Moffatt also participated in the Malibu studies. These consultants used some proprietary data supplied by Ford and General Motors in their studies.

On the other hand, there were researchers from the Insurance Institute for Highway Safety (IIHS), the insurance advocacy group. Part of the mission of IIHS is to promote safety on our roadways. Their studies reinforced NHTSA's position.

The two rival groups came to opposite conclusions about the importance of roof strength. Yet, they used similar datasets.

Both groups of researchers used police accident reports from states across the country. In contrast to the detailed information on rollover crashes in NHTSA's NASS-CDS data, the state datasets were very limited. Police simply did not record certain data. For example, the number of quarter turns in a crash was not in the police accident reports. Therefore, the researchers could not eliminate rollovers with only a 1 quarter turn. Nor could they include the number of quarter turns in their analysis. Furthermore, they could not identify or remove rollovers that had been halted by a tree, wall, or fence. Moreover, the analysts could not determine the source of the occupant injury. The state data files did not contain information about how the occupant was injured or killed. In total, these police accident reports lacked details about the rollovers and the occupant injuries.

Because the accident reports did include the number of vehicles in a crash, the investigators could follow NHTSA's approach and select only single-vehicle-rollover crashes with front-seated, (mostly driver) occupants.

However, unlike the federal research, both groups supplemented their crash data. They added information about the roof strength of the vehicle. With this added data, the analysts could search for a link between roof strength and occupant injury. They used these additional measurements to test NHTSA's theory. If NHTSA was right, then weak roofs would lead to more injuries and stronger roofs to fewer injuries.

What did they find? Short answer, different results:

- **The industry consultants found that roof strength had no effect on the risk of occupant injuries.**

- **The IIHS researchers found that stronger roofs significantly decreased the risk of occupant injuries.**

If both groups were using police accident reports, then how did they arrive at different conclusions?

The different results were primarily due to the factors that each group chose to include or exclude in their analysis, as highlighted in Figure 5.7. For example, the industry consultants used alcohol information from the police accident reports. Based on this analysis, drunk drivers had rollovers with more severe injuries regardless of the roof strength. Similarly, these consultants used seat belt status in their analysis. Their conclusion was unbelted occupants had more severe injuries than belted occupants, regardless of roof strength.

In contrast, the IIHS researchers deemed these data unreliable and unusable. Police officers frequently made a subjective evaluation about whether or not a driver had been drinking. In other words, the information was not based on a scientific test. Secondly, seat belt information was

OPPOSING		SUPPORTING	
Alcohol testing results or police opinion about alcohol usage should be used in analysis.	Whether or not occupant is belted influences injury.	People lie about belt use to police; information is unreliable.	Alcohol testing is not consistent; police opinion is unreliable.
	Even when some of the data are miscoded, they are still useful.	Injury coding by police is unreliable; it is better to only use occupants with fatal injuries in analysis.	
	Serious injuries, as well as fatalities, should be included in the analysis.		
		Stronger roofs reduce occupant ejection risk.	
	Occupant ejection is unrelated to roof strength.		
Our data are more representative; your data are not.			
		Separate analysis of SUVs and passenger vehicles is appropriate.	
Vehicle's aspect ratio should be included in analysis.	Where a crash occurs, i.e., urban or rural road, should be included in analysis.	Vehicle's static stability factor is a better factor to consider than a vehicle's aspect ratio.	Where a crash occurs, i.e., urban or rural road, does not change the analysis results.

FIGURE 5.7   Highlights of data-related issues in the debate. Industry consultants and researchers at the Insurance Institute for Highway Safety chose different factors in their analyses leading to different conclusions.

unreliable because it was often self-reported by the occupant. And some occupants lie.

After criticism from the industry consultants, the IIHS researchers did run additional analyses using only belted occupants. Their conclusion, across all their analyses, remained the same. Roof strength was significantly related to occupant injury. Weaker roofs were associated with more severe injuries.

The two groups also differed on what vehicle attributes to include. For instance, the industry consultants thought that the aspect ratio of a vehicle was a useful predictor of injury and should be included. The aspect ratio is the vehicle's height divided by its track width. Cars with low aspect ratios tend to be shorter and therefore have less room above the occupant's head. In theory, this could have been an important factor in occupant injuries in rollovers. But in the end, the aspect ratio was not found to be a good predictor of occupant injuries.

The researchers at the IIHS instead used the vehicle's static stability factor (SSF) in their analysis. The SSF is defined as half the vehicle's track width divided by the height of the vehicle's center of gravity. Vehicles with low values have a greater tendency to roll over than vehicles with higher SSF values. These researchers considered the SSF a reasonable surrogate for rollover severity because low-value SSF vehicles also would tend to have more rolls and, possibly, more severe injuries. In the end, the SSF was not found to be a good predictor of occupant injuries.

Furthermore, the IIHS researchers did two separate analyses on different types of vehicles. One analysis only included SUVs. The other, only passenger cars. Again, their conclusions did not change.

While both groups started with police accident reports, each group analyzed very different sets of data in the end. It is not surprising that they arrived at contrary results.

THE DECISION AND AFTERMATH

Finally, after reviewing these statistical studies and hundreds of comments submitted to the docket over 4+ years, the NHTSA administrators made a decision.

On May 12, 2009, NHTSA issued its final rule for upgrading the Roof Crush Resistance Standard. In addressing the statistical debates, NHTSA analysts did not condemn the studies performed by the industry consultants, nor condone those studies issued by the IIHS. Both sets of studies presented some insights.

However, NHTSA made it clear that it considered its own research as providing the best answer to the question about the role of roof strength in occupant injury. Only the NASS-CDS data identified the source of the occupant injuries. Therefore, NHTSA researchers were confident that their analysis demonstrated a strong link between roof intrusion and injury.

In reaching its decision, NHTSA considered the additional argument made by the industry about the cost of a new standard. Carmakers would incur an extra cost when they redesigned vehicles to have heavier roofs. And they would pass that cost on to car buyers. Plus, there would be added fuel costs for ownership because it takes more fuel to move heavier vehicles. Even so, the federal government decided that these costs paled in comparison to the benefit of preventing an estimated 135 fatalities and 1,065 nonfatal injuries each year.

Going forward, cars would be tested at a roof strength 3.0 times the unloaded vehicle weight; light trucks would be tested at 1.5. The upgraded standard, called FMVSS 216a, was phased in through model year 2015.

In parallel, the IIHS introduced a safety rating system for vehicle roof strength as part of its consumer information campaign for the "Top Safety Pick." The purpose was to incentivize manufacturers to build even stronger roofs. As you will read in the chapter on vehicle safety ratings, manufacturers are now building roofs that exceed the national standard in order to be awarded the IIHS's highest safety award.

But are these stronger roofs reducing the severity of occupant injuries?

In November 2020, NHTSA published a study that evaluated the effectiveness of stronger roofs. The new analysis used nearly 2,000 occupants in single-vehicle-rollover crashes from state police accident reports. The results revealed that occupants in vehicles with stronger roofs (after the upgrade to the standard) had decreased risk of sustaining a severe injury in a rollover crash. The risk was reduced by 20%. The data showed that stronger roofs were better.

NOTE FROM THE AUTHOR

The analyses described in this chapter serve as a prime example of the role that data play in a regulatory debate. There are three issues that I believe deserve further discussion.

The first issue concerns using the wrong data to study a problem. In my opinion, the state data fall into this category. These data are wrong because (a) they are unreliable and (b) lack important details about the crashes.

I already talked about the unreliability of the seat belt usage data. But there is also a reliability issue concerning the injury data. Research shows that police often, as much as 60%, code occupant injuries as more severe than medical personnel do. For example, what a police officer may deem as a severe or incapacitating injury, perhaps due to the amount of blood at the scene, actually turns out to be only a minor injury. When analyses depend on injury data, as the industrial studies described in this chapter do as well as the NHTSA study in 2020, then their results are unreliable. Unreliable data lead to unreliable results.

Furthermore, the state data do not contain important information for studying the effect of roof strength on injuries. As I described earlier, when a rollover vehicle only experiences a 1 quarter turn, then there is no potential roof-to-ground contact. Researchers estimate that about 25% of rollover crashes only have a 1 quarter turn. The federal analysts eliminated these crashes from the NASS-CDS datasets. That dataset had the details needed to do so. But the analysts using the data from police accident reports could not. They did not have the details in the dataset. They could not identify 25% of the rollovers that had no roof-to-ground contact. Hence, 25% of their data were irrelevant to the problem of roof strength analysis. As a result, I believe that these analyses used the wrong data.

The second issue addresses the strength of the conclusion at the end of an analysis. In the roof crush resistance debate, the NHTSA researchers built statistical models to determine the link between what happened to a vehicle in a crash and what happened to the occupant. The NASS-CDS data contained details about the amount of roof intrusion, the source of the occupant injuries, and severity of the injuries. Unlike the police accident data, the sole purpose of the NASS-CDS data collection is to obtain reliable, in-depth information useful for studying potential safety problems. The nationwide, representative sample is considered very high quality. A comprehensive analysis with high-quality data produces strong conclusions.

The third issue concerns causality. All crash data are observational data, that is, information collected after the occurrence of a crash. Unlike experimental data in a laboratory setting, observational data limit an analyst's ability to clearly point to a cause-and-effect. While the data used by NHTSA identified the source of the injury, the analysts could not *definitively* say that roof intrusion **caused** the injury. But they could show that, statistically, more intrusion was **strongly linked** to higher fatal and serious injury occurrence. Herein lies a key criticism voiced by industry representatives. In their opinion, NHTSA could never **prove** that the roof intrusion caused the injury.

For some, that statistical link is not good enough. For me, it is.

NOTE

1. Graham, John D. (1989). *Auto safety: Assessing America's performance.* Auburn House Publishing Company, p. 236.

CHAPTER 6

The Incompatibility Debate

"You don't have to worry about me, Mom. I'm bigger than they are." This was a 16-year-old boy's suggestion to his mother as he left home, driving his family's Chevrolet Suburban.

—*DIANA CALICA (1996)*[1]

Sometimes, lessons in car safety come from unexpected places.

Something was missing when National Hockey League (NHL) player Zack Smith got his equipment at his first development camp: a visor. He was the only player not to receive one and Smith never went back to ask for one. Without eye and nose protection, Smith says he got a reputation for being extra tough. Truth is, he was scared.

A study published in the *Journal of Public Economics* suggests that Smith may have had good reason to be wary. Its authors used the 2004–2005 NHL labor lockout to show that players tend to behave more aggressively when they are forced to wear visors.

When the NHL cancelled its season, hundreds of players went to skate in the European leagues. Unlike the NHL at the time, the European teams required visors. That gave the study's researchers

DOI: 10.1201/9781003035343-6

a perfect natural experiment. They could compare players' behavior with and without visors while controlling for other differences between the leagues.

The researchers found that the players who did not normally wear visors, but were forced to, the compliers, played more aggressively in the European games than did the always wearers. In fact, the compliers earned an additional 0.4 penalty in minutes (PIM) per game – significant since the average PIM per game is 1.13. In other words, when forced to wear visors, these players were more reckless.

The phenomenon of taking more risk, with the use of added personal safety equipment, may have a direct and dangerous parallel in car safety – one that you may have experienced personally.

During the regulatory debates of the 1970s, an economist from the University of Chicago, Samuel Peltzman, published a paper that argued against seat belt mandates. The fundamental idea was that: *With an increased sense of personal safety, humans take more risk that could harm themselves and others.*

Peltzman believed that installing seat belts and mandating their use would not yield an overall benefit. Using statistical models, he predicted that drivers, feeling better protected, would take more risks. The expected change in behavior would, in turn, result in more accidents. Therefore, the unintended consequences of seat belts would be more deaths, not less.

Peltzman's idea stirred up a debate. Car safety advocates and other economists disputed his data, analysis, and conclusions. Nearly 50 years later, do we know who was right?

According to National Highway Traffic Safety Administration's (NHTSA) research, seat belts have been quite effective in saving lives. Hundreds of thousands of lives have been saved since the early 1960s. Today, there is little doubt that seat belts are an important, positive safety device. Peltzman was wrong about seat belts.

Nonetheless, the fundamental idea is still under debate in some circles. In fact, this notion of more reckless behavior with increased personal protection is popularly called the Peltzman effect. Others call it the Gladiator effect.

In 2007, a study examined National Association for Stock Car Auto Racing (NASCAR) drivers before and after their cars were outfitted with new safety devices. The researchers found that the new equipment led to more accidents during races, but fewer injuries. (And since more accidents increased viewership, NASCAR deemed the safety devices a product marketing success.) In short, both the NASCAR and the hockey studies came to the same conclusion: More safety devices led to increased risk taking.

Does this Gladiator effect have any merit when discussing car safety on our nation's roadways? What if car safety researchers have largely discounted this effect because of a lack of the right kind of data? What if this behavior can only be identified using data of individual behavior before and after safety equipment is installed?

After all, you likely have experienced a related phenomenon. Ever sit in a sedan at a traffic light and look up at the driver of a pickup idling alongside you? In your gut, you know that your sedan would lose in a crash. And you are right, for the most part.

On the other hand, how do drivers of larger trucks feel when they look at your sedan? Perhaps these drivers behave differently because they feel invincible. Do they have good reason to feel this way?

The short answer is "yes." People in smaller, lighter cars are more likely to be injured or killed than people in larger pickup trucks. And a lot of it comes down to physics.

THE INCOMPATIBILITY PROBLEM

Car safety experts refer to this uneven occupant risk of injury or death as **vehicle incompatibility**. Many factors influence the degree to which two vehicles are incompatible, including differences in the structural stiffness and shape of the vehicles. For example, the front of a light truck may be more rigid than the car it hits. Or, their bumpers are not at the same level. In both examples, the light truck could impart a force that the car is not designed to withstand. As a result, car occupants experience worse outcomes.

Stiffness and shape are important factors in determining the degree of incompatibility, but the most critical factor is weight. In a crash, the occupants of the lighter vehicle, due to physics, face a higher risk of injury. Figure 6.1 of a frontal collision helps to explain this concept.

Prior to the crash, both vehicles are moving. At impact, both vehicles experience the same forces. But, because of the physics related to the conservation of momentum, the post-crash acceleration of each vehicle will be different depending on its weight. When two vehicles are unequal

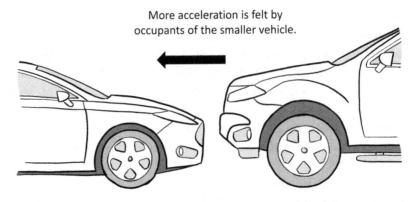

More acceleration is felt by
occupants of the smaller vehicle.

FIGURE 6.1   Illustration of a two-vehicle, frontal collision between a car and a light truck. The incompatibility problem refers to the uneven risk of injury or death of vehicle occupants. The occupants in the smaller vehicle are at higher risk.

in weight, more of the energy from the collision goes into the lighter vehicle. More energy also means more acceleration of the occupants' bodies in the lighter vehicle, forcing them forward. The bigger the difference in the vehicle weights, the more uneven the energy distribution is between the two vehicles. Despite modern safety equipment such as seat belts and airbags, these occupants have a higher risk of injury. The bigger the difference in vehicle weights, the higher the risk of the occupants in the lighter vehicle.

Researchers typically use vehicle **curb weight** to study this problem. Curb weight is the weight of the vehicle with a full tank of gas and all fluids, but without occupants or cargo. Naturally, curb weight underestimates the weight of a vehicle at the time of a crash because the weight of the driver is not included. Not to mention any additional occupants or cargo. Nonetheless, curb weight is nearly always used for these types of analyses concerning incompatibility.

Since weight is so important, it helps to have some context. Table 6.1 lists the curb weight of some popular vehicles for model year 2020. Each vehicle, depending on the features and options chosen (e.g., engine size, transmission, size of passenger compartment in a pickup truck), has a range of weight values. Incompatibility can exist within classes of vehicles, but as you see in Table 6.1, the weight difference is usually not as great.

The graphical display in Figure 6.2 highlights the substantial differences in curb weight. For example, the lightest Toyota Prius (vehicle #5) at

TABLE 6.1   Curb Weights of Some 2020 Model Year Cars and Light Trucks. The range of weights for a vehicle is a result of different options, such as engine size or transmission.

Vehicles		Curb Weight (lb)	
Cars		Low	High
SmartCar forTwo	1	1,940	1,940
Honda Civic	2	2,762	3,009
Hyundai Electra	3	2,844	3,131
Toyota Corolla	4	2,910	3,150
Toyota Prius	5	3,010	3,220
Nissan Altima	6	3,208	3,461
Ford Fusion	7	3,241	3,572
BMW 325i	8	3,589	3,589
Light Trucks			
Subaru Forester	9	3,449	3,589
Volvo XC40	10	3,574	3,805
Toyota Tacoma	11	3,980	4,480
Honda Pilot	12	4,036	4,319
Ford F150	13	4,069	5,684
Lexus RX 350	14	4,222	4,387
Chrysler Grand Caravan	15	4,321	4,483
Ford Explorer	16	4,345	4,727
Chevrolet Silverado 1500	17	4,520	5,240
Jeep Gladiator	18	4,650	5,050
Chevrolet Suburban	19	5,586	5,808
Dodge Ram 2500	20	5,988	7,431

3,010 pounds and the heaviest Jeep Gladiator (vehicle #18) at 5,050 pounds has a weight difference of over 2,000 pounds. Even among more popular vehicles, the weight difference could be over 1,000 pounds.

These few examples illustrate a pattern. Researchers at the Insurance Institute of Highway Safety looked into this and found that during the years 2013–2016, the average SUV outweighed the average car by 1,000 pounds. Pickups were 2,000 pounds heavier.

Neither the federal government nor the industry have taken steps to restrict the size and weight of vehicles sold. Instead, there is some movement to mitigate the effect of the differences. That is, manufacturers need to make heavier trucks **less aggressive**, i.e., less harmful to others, and make lighter vehicles **more crashworthy**, i.e., improving the protection of occupants in smaller vehicles. How is that going?

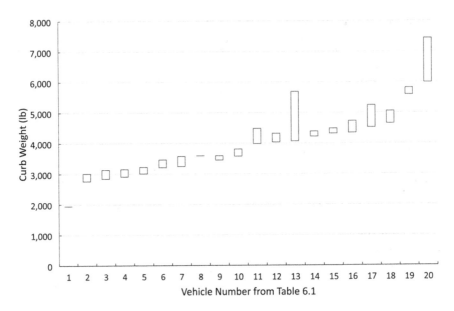

FIGURE 6.2 The high & low curb weights of 2020 model year cars and light trucks listed in Table 6.1. The graphical display of the 20 vehicles highlights the large differences in weights among the vehicles, especially cars (vehicles 1 through 8) and light trucks (vehicles 9 through 20).

MEASURING THE INCOMPATIBILITY PROBLEM

In order to measure progress, there first needs to be a way to measure the problem. In a 1996 publication, NHTSA analysts identified the "Top 20 Aggressor" vehicles. They looked at two-vehicle crashes and determined who was harmed. The researchers labeled each crash vehicle as either the "subject vehicle" (potential aggressor) or "other vehicle."

For their measurement system, they used data from 1991 to 1993 and defined the

Aggressivity Metric of Subject Vehicle

$$= \frac{Number\ of\ Deaths\ in\ Other\ Vehicles\ in\ Crashes\ of\ Two\ Vehicles}{Total\ Registration\ of\ Subject\ Vehicle\ (in\ millions)}$$

The metric counts the fatalities in the other vehicles in collisions with the subject vehicle. It normalizes the fatality count by the number of subject vehicles on the road. That is, a vehicle's aggression is measured by the amount of harm caused to occupants of other vehicles, scaled by the total number of subject vehicles on the road.

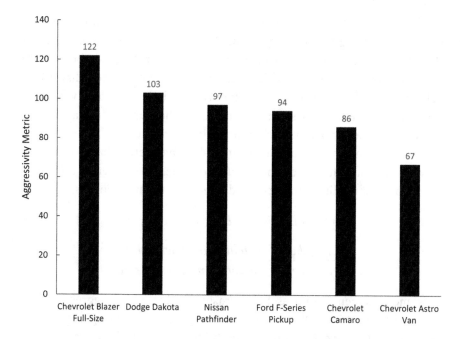

FIGURE 6.3   Some of the top 20 aggressors in 1991–1993. The higher the number, the greater harm inflicted on occupants of the other vehicle in a crash.

For example, these researchers reported that the Ford F-series pickup was involved in 2,474 fatal, two-vehicle crashes. The crashes resulted in 606 fatalities among the pickup occupants and 2,341 deaths of people in the other vehicles. There were about 24.9 million of these pickups on the road (or about 8.3 million a year) during that time period. Therefore,

$$Aggressivity\ Metric\ of\ Ford\ F-Series\ Pickup = {}^{2,341}\!/_{24.9} = 94.$$

How does this compare to other vehicles?

Figure 6.3 illustrates some of the researchers' findings. All but one of the top 20 aggressive vehicles were larger, full-size pickup trucks and vans. The most aggressive vehicle was the Chevrolet Blazer, a full-size sport utility vehicle with an aggressive metric of 122. It had an estimated curb weight of 4,700 and a structural stiffness designed for off-road use. This supported NHTSA's thinking that both the weight of the vehicle (in comparison to other vehicles on the road) and its design features (such as frame stiffness) influenced aggression. The Ford F-Series pickup, with an aggressivity metric of 94, ranked 7th on the longer list in the original study.

The only car that made the list was the Chevrolet Camaro with a metric of 86, ranking the 9th highest in the original study. The Camaro at that time was a mid-size, performance vehicle that weighed about 3,200 pounds. The researchers noted that its ranking demonstrated that driver behavior may have played an important role in these calculated values. In other words, the Camaro's ranking may be higher due to the way the car was driven – rather than its weight or design attributes.

The researchers then introduced an alternative metric that reduced the driver behavior influence on the calculated value. In this metric, they used the number of fatalities that occurred in each of the two vehicles:

$$\textit{Fatality Ratio of a Subject Vehicle}$$

$$= \frac{\textit{Number of Fatalities in Other Vehicles}}{\textit{Number of Fatalities in Subject Vehicle}}$$

This measurement captures both crashworthiness and aggressiveness. That is, how protective the other vehicles are relative to the aggressiveness of the subject vehicle. If this ratio is high, then either (1) the subject vehicle inflicts more harm on the occupants of the other vehicle while protecting its own occupants and/or (2) the other vehicles are not sufficiently crashworthy when involved in a crash with the subject vehicle. An example will help to clarify these notions.

Using the Ford F-Series pickup, with 606 fatalities in the pickup and 2,341 in the other vehicles:

$$\textit{Fatality Ratio of the Ford F-Series Pickup} = \frac{2,341}{606} = 3.9.$$

That is, occupants of the other vehicles that crashed with a Ford F-Series pickup died at nearly 4 times the rate of occupants in the pickup. Such a value indicates that the pickup was aggressive (it caused harm to others and better protected its own occupants) and/or other vehicles were not sufficiently crashworthy (they did not protect their occupants against this pickup).

Figure 6.4 displays the fatality ratio of some of the subject vehicles reported in the study. The Camaro now ranks lower with a value of 1.34. To interpret these ratios, here are some highlights:

- When the ratio is greater than 1, this shows that more occupants died in the other vehicle than the subject vehicle. For example, when in a

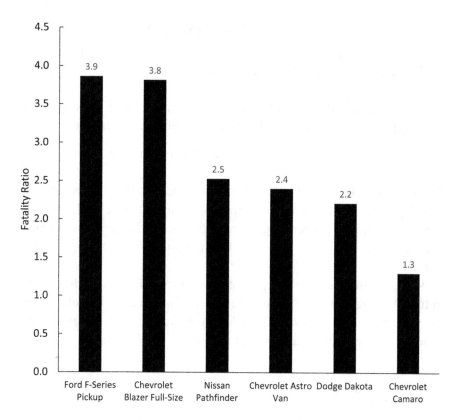

FIGURE 6.4 Fatality ratios of some of the top 20 aggressors in 1991–1993.

two-vehicle crash with a Ford F-series pickup, occupants in the other vehicle died 3.9 times more often than the pickup occupants.

- On the other hand, the Camaro's ratio of 1.34 is relatively close to 1. Yes, it is greater than 1 and therefore implies that more occupants died in the other vehicle. However, it also indicates that the Camaro did not do a good job of protecting its own occupants.

Published in 1996, these measurements provided the statistical evidence needed to identify the problem of incompatibility on our roadways. Many unmeasured factors influenced these numbers, such as the lack of restraint use, the type of crash (frontal versus side-impact), and age of the impacted vehicles. Nonetheless, NHTSA decided to address the problem.

THE PROPOSED REMEDY

In June 2003, NHTSA opened a public docket to fix the incompatibility problem. The agency wanted manufacturers:

1. To change the design of light trucks to make them less aggressive (while not compromising the safety of their occupants) and

2. To increase the protection of car occupants in a crash.

NHTSA used the data of fatalities in two-vehicle crashes involving a car and light truck, as shown in Figure 6.5, to demonstrate an increasing trend that needed correction. More people were dying in two-vehicle crashes involving a car and light truck than any other type of two-vehicle crash.

Plus, researchers foresaw car occupants facing increased risk in the future. New light-truck purchases were on the rise. In 2001, half of all vehicle purchases were light trucks, including pickups, SUVs, and minivans. And, as discussed in Chapter 4, that percentage was increasing.

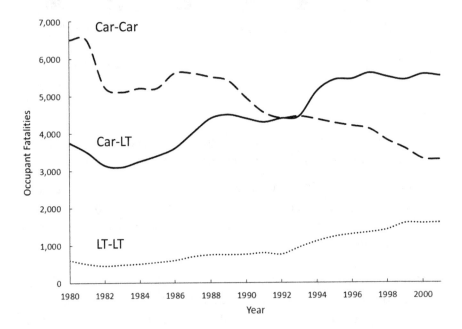

FIGURE 6.5 Trend in number of all occupant fatalities in two-vehicle crashes of cars and light trucks (LT), 1980–2001. Light trucks include pickups, SUVs, and minivans.

NHTSA dug deeper into the data to support their cause. Analysts wanted to draw attention to the most harmful crashes involving a car and light truck:

- When a light truck side-impacted a car, the car occupants died 30 times more often than light-truck occupants and

- In frontal crashes, car occupant died 4 times more often than light-truck occupants.

Given these comparisons, the problem was clearly identified. Something had to be done to improve the compatibility of light trucks and cars.

When NHTSA opened the docket in 2003, it was looking for a solution, either by creating a new standard that carmakers must conform to or by finding some other novel approach. Industry leaders met with NHTSA officials and ironed out an approach that was intended to address both the lack of crashworthiness of cars and the aggressivity of light trucks. The result was a voluntary commitment by 15 manufacturers issued in December 2003. The participants were BMW Group, DaimlerChrysler Corporation, Ford Motor Company, General Motors, Honda, Hyundai Motor, Isuzu Motors, Kia Motors, Mazda, Mitsubishi, Nissan, Subaru, Suzuki, Toyota, and Volkswagen Group.

To address the lack of crashworthiness of cars, the manufacturers committed to install better protection. This led to the widespread adoption of side and curtain airbags, stronger side panels, and better padding in doors.

To decrease the aggressivity of light trucks, the makers would redesign the front structure of these vehicles to correct for the mismatch of the energy-absorbing structures, typically behind the bumper. To align these structures to reduce the forces on occupants in a crash, manufacturers offered two possible solutions:

- Option 1: Lower a truck's primary, energy-absorbing structure to adequately overlap the that of passenger cars;

- Option 2: Add a secondary, energy-absorbing structure, frequently called a blocker beam, below the primary structure of the light truck to align with passenger cars. This option was intended to suit vehicles that sit higher off the ground.

The commitment had some built-in, rather vague exceptions. Vehicles that were "specifically designed primarily for off-road use" and vehicles that the "manufacturer determines cannot meet the performance criteria without severely compromising their practicality or functionality" were exempt from this commitment.

Manufacturers agreed that all relevant light trucks would conform to these requirements by September 2009. However, progress appeared much faster with about half of light trucks meeting this voluntary design specification by 2004. Has this changed the trend seen in Figure 6.5?

MEASURING PROGRESS

The 2003 Industry Commitment appears to have had some impact with a decreasing fatality trend between 2005 and 2010, as shown in Figure 6.6. However, after 2014 the fatality trend moved upward for occupants in car-light truck crashes, as has the trend of other two-vehicle crashes.

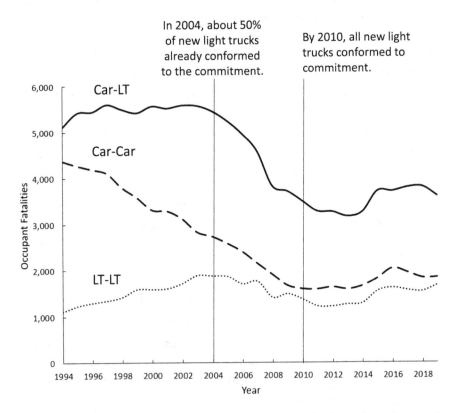

FIGURE 6.6 Number of all occupant fatalities in two-vehicle crashes of cars and light trucks (LT includes pickups, SUVs and minivans) before and after 2003 industry commitment.

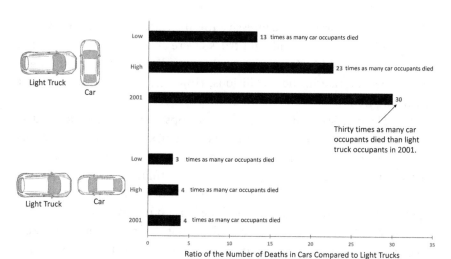

FIGURE 6.7 The high and low annual fatality ratios of two-vehicle collisions of a car and light truck during the period 2009–2018. The 2001 numbers from NHTSA's 2003 docket serve as a comparison to measure the progress made in reducing fatalities in cars.

NHTSA analysts examined the car-light truck fatalities more closely and issued a progress report in 2012, some nine years after the commitment. The results were mixed. On the one hand, pickup trucks had gotten slightly more aggressive. On the other hand, SUVs had become slightly less aggressive.

As discussed earlier in this chapter, NHTSA used fatality ratios to draw attention to the most severe crashes involving a car and light truck. Do the data still show a problem of incompatibility?

Figure 6.7 gives the comparison of the worse and best annual fatality ratios over a ten-year period, 2009–2018:

- When a light truck side-impacted a car, car occupants died at a rate of 30 times more often than light-truck occupants in 2001. In the past ten years, the worst annual rate was 23 and the best rate was 13.

- In frontal crashes, car occupant died at a rate of 4 times more often than light-truck occupants in 2001. In the past ten years, the worst annual rate was 4 and the best rate was 3.

While the problem appears to have lessened, it has not disappeared.

In 2019, the Insurance Institute for Highway Safety (IIHS) published another, more detailed assessment. They found some progress. Pickup trucks and SUVs were still aggressive toward cars, but to a lesser degree. In particular, when cars and light trucks were closer in weight, then they were more compatible. Occupants in cars and light trucks were dying at about the same rate in these collisions. However, the analysts also saw a disturbing trend.

The IIHS analysts found that modern pickups are getting heavier relative to cars. In the late 1980s, the average pickup truck was about 30% heavier than the average car in the two-vehicle crashes studied. But recently, the average pickup truck was nearly 100% heavier than the average car. That is, these pickups were nearly twice the weight of the average car. And as expected, these heavier vehicles posed the highest risk to occupants of cars.

This weight inflation in trucks does not appear to have peaked. According to forecasted sales, larger and heavier vehicles are increasing in popularity. Consequently, the weight gap between passenger cars and trucks, in particular pickup trucks, appears to be growing. And with it, the problem of incompatibility.

NOTE FROM THE AUTHOR

I began this chapter with a discussion of the Peltzman or Gladiator effect. That is, people take more risks when they feel more protected. Currently, the exact type of data needed to prove that this effect applies to drivers of heavy pickup trucks, SUVs, and minivans does not exist. However, there is no disputing that the larger vehicles provide more protection. Furthermore, there is clearly a subset of the driving population that enjoy taking risks. It does not require a quantum leap to conclude that when these risk-taking individuals choose to drive these oversized vehicles there is an increased hazard to the rest of society. Bigger vehicles mean bigger problems for those sharing the road.

However, there is also another factor that is more heart-wrenching for all concerned. Simple human error.

> "Like a monster truck driving over vehicles in a show." That is how one person described the car crash that killed a 3-year-old girl. She was secured in a child-safety seat in the second row of her family's compact sedan, a 2019 Toyota Corolla.

> The little girl's mother was driving them down a rural, two-lane road in northeast Arizona. It was a clear Saturday afternoon, and they were headed to a family event nearby.

As she later told the police, this young mother slowed the Corolla to let an oncoming car pass so she could turn left onto a side road. She checked her mirrors and saw a 2018 GMC Sierra (full-size, pickup truck) approach from behind.

The truck's driver said she glanced down at her vehicle's infotainment center for just a moment and hit the brakes when she saw the sedan.

Even so, the 4-wheel-drive truck hit the back of the sedan with enough force to crush through to the C pillar. That's the roof support structure on either side of the rear window.

A police photo of the sedan shows an elevated and mangled trunk with its contents spilling out. In contrast, the truck had a broken headlamp and some moderate damage to its grill.

Both of these vehicles were built after the 2003 Industry Commitment. Yet, the GMC pickup was too aggressive for the rear-seated child in the Corolla. And the industry has known for many years about how to reduce the number of these tragic events.

Driver-assist technology for preventing crashes, such as forward-collision warning, has been available for more than a decade. As I will discuss in Chapter 10, the evidence is overwhelming that these technologies can reduce crashes by as much as 50%. Yet, they did not become standard equipment on new vehicles until 2022.

In my opinion, these technologies should have been mandatory long ago, especially on the most aggressive vehicles, like the 2018 GMC Sierra. Occupants in "other" vehicles need to be protected from these larger vehicles. If not by vehicle design, then by preventing the crash, altogether. While the industry cannot change the laws of physics, it has known for too long how it can influence human behavior and reduce the catastrophic effects of human error.

NOTE

1. Interview on March 15, 2021 with the mother, Dr. Diana Calica, about her experience in raising her young son.

CHAPTER 7

The Fuel Efficiency Debate

Specifically, over the last 20 years – and perhaps particularly over the last decade – there has been a steady increase in the attention that car buyers pay to safety concerns …. This is germane to the committee's work because the possible effects on safety of the original CAFE program, as well as the effects on safety that a renewed effort to improve fuel economy would have, have been perhaps the most controversial aspect of the program.

—NATIONAL RESEARCH COUNCIL (2002)[1]

The nation received a wake-up call in 1973. The Arab oil-producing countries imposed an embargo on the United States after its support of Israel in the Yom Kippur War. Long lines at gas stations signaled that our economy was vulnerable.

In response, Congress passed the Environmental Policy and Conservation Act of 1975. The intent was to make our country less dependent on foreign fuel. The Corporate Average Fuel Economy (CAFE) program was part of this act. This program required manufacturers of cars and light trucks to increase the fuel economy of vehicles sold in the United States. The act set a national goal of doubling the fuel efficiency of new passenger vehicles to 27.5 miles per gallon (mpg) by 1985. The National Highway Traffic Safety Administration (NHTSA) now had the authority to set and enforce this standard. Table 7.1 shows the early standards set for cars leading up to the 1985 target.

DOI: 10.1201/9781003035343-7

TABLE 7.1   Corporate Average Fuel Economy
Standards (CAFE) 1978–1985. Congress set the
first target for increasing fuel efficiency in 1975.

Year	Average Fuel Economy Standard (Miles per Gallon)
1978	18.0
1979	19.0
1980	20.0
1981	22.0
1982	24.0
1983	26.0
1984	27.0
1985	27.5

Early on, many carmakers complied with the standard by making
certain cars in their fleets smaller and lighter. Dr. Leonard Evans was a
General Motors researcher known for his love of data. In the early 1980s,
he published two papers that said making cars lighter would kill people.
Using fatal crash data, he developed curves, as illustrated in Figure 7.1, that
showed how vehicle weight was tied to fatality risk.

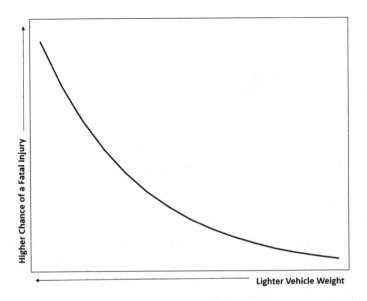

FIGURE 7.1   Illustration of relationship between chances of a fatal injury in a
crash and vehicle weight. As vehicles got lighter, the chances of fatal injuries in
crashes increased.

This finding was not new. As I discussed in Chapter 6, bigger vehicles typically protect their occupants better and cause harm to others in a crash. In the 1980s, the car safety community understood this. However, Evans' work went further. He gave specific curves, similar to Figure 7.1, that could be used to estimate how much weight meant life or death.

STRUCTURE OF THE EARLY CAFE STANDARD

Unlike the Federal Motor Vehicle Safety Standards discussed in this book, the CAFE standard is not an individual vehicle standard. Rather, it is a fleet-wide standard for each manufacturer. The fuel economy target must be met, collectively and on average, for all the new vehicles sold by a car-maker in a given year.

For example, in 1978 the standard was 18 mpg. If Ford widely sold its new, full-size sedan LTD model that got 17 mpg, then the company also needed to widely sell its more fuel-efficient Mustang that got 26 mpg to balance their sales. Less fuel-efficient vehicles, so-called gas guzzlers, had to be balanced by selling more fuel-efficient cars. At the end of the year, NHTSA would compute a composite average of Ford's sales to determine if the company met the standard of 18 mpg.

The composite average, more precisely the harmonic average, accounts for the vehicle's sales volume and mpg. As illustrated in Figure 7.2a, with the stacked vehicles representing sales volume, the mpg fulcrum is intended to show how the manufacturers had to balance their sales to meet the standard. Hence the name, Corporate *Average* Fuel Economy.

There are many market factors that influence car sales. To adjust for these uncertainties, the standard also has a system of penalties and credits. As illustrated in Figure 7.2b, if the company's composite average mpg is too low, then the company is subject to a financial penalty. The poorer the mpg (i.e., too low), the more the company has to pay the federal government.

However, if the company generates a composite mpg that is higher than the legal target, then this is a good thing. The company is selling vehicles with better mpg, on average, than required. Hence, the company earns credits that they can use to help balance any future penalties.

While NHTSA sets the mpg target, each company is free to choose a different strategy to meet the requirement. Strategies might include discontinuing gas guzzlers or designing more fuel-efficient models. Early on, many carmakers reduced the weight of vehicles they wanted to retain to satisfy the CAFE standard. Lighter cars require less fuel than heavier ones to go a mile.

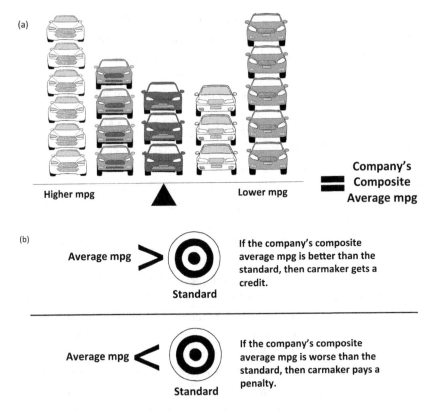

FIGURE 7.2 (a) Illustration of a company balancing vehicle production to meet the first CAFE standard. (b) Illustration of the credits and penalties of the first CAFE standard.

IS THERE A SAFETY TAX?

Economists Dr. Robert Crandall from the Brookings Institute and Dr. John Graham from the Harvard School of Public Health published a paper in 1989 that warned the CAFE standards had a safety tax. Using Evans' equations and some additional analysis, these researchers concluded that the CAFE standards would create a public health issue. More people would die as a result of manufacturers making cars smaller and lighter. Crandall and Graham wrote:

> Our rough estimate is therefore that the 500-pound or 14 percent reduction in the average weight of 1989 cars caused by CAFE is associated with a 14-27 percent increase in occupant fatality risk. ...
> We estimate that these 1989 model year cars *will be responsible* [emphasis added] for 2,200-3,900 additional fatalities over the next 10 years because of CAFE.

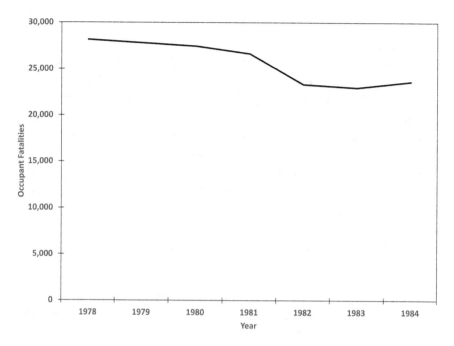

FIGURE 7.3   Car occupant fatalities between 1978 and 1984. During this time period, the average weight of new cars dropped by about 500 pounds.

Federal analysts disagreed and pointed to fatality statistics. Between 1978 and 1984, car occupant fatalities decreased nearly 20%, as shown in Figure 7.3. At the same time, the average weight of new cars decreased by about 500 pounds and vehicles became more fuel efficient. How could anyone look at this data and say that there was a safety tax associated with the CAFE standards?

Crandall and Graham said there were more factors at play. People drove less during the recessions of 1980 and 1982, and they were less likely to drive drunk because of a national awareness campaign. Furthermore, it was natural to expect fewer deaths because newer cars had more safety features and more people were able to buy these cars because of rising incomes. Nonetheless, these researchers insisted in their paper that the CAFE standards had a safety tax:

> We submit that the decline in car occupant fatalities from 1980 to 1985 *might have been more dramatic had CAFE not been in effect* [emphasis added].

Notice the phrase "might have been." This idea of looking back and guessing (albeit based on reasonable scientific formulas) about what "might have been" is one of the problems facing statisticians who study the effects of the CAFE standard. For example, suppose a driver died in a 1984 sedan that weighed 2,500 pounds. Would he have lived if the vehicle was 100 pounds heavier? Analysts in the 1980s, similar to Evans, studied ways to answer the "what might have been" question. And the more they learned, the more complex the problem became.

Before the early 1990s, most analysts in the industry recognized the possible risk of making vehicles lighter and smaller. However, there were also possible benefits, such as:

- If all vehicles were lighter and smaller, then multi-vehicle collisions would be less severe.

- Lighter and smaller vehicles (perhaps due to driver-car interaction) could better maneuver around cyclists and hazards on the side of the road, like trees and fences.

- Since lighter and smaller vehicles require less horsepower to move, then their impacts would be less severe.

- Lighter and smaller vehicles are less prone to rolling over since their center of gravity is lower and wheelbase wider.

After studying nearly 20 statistical studies on this topic, a committee of the National Research Council concluded that there was no clear answer to the safety question. They summarized their thoughts in their 1992 report:

> Available [statistical] studies arrive at conclusions that often appear contradictory. Further, because of the difference in the data examined and the analyses conducted, most studies are not directly comparable. In short, although the relationship of fuel economy to safety is of central interest, the issue is surrounded by substantial confusion.

Amid this debate, the nation moved toward requiring more fuel economy in cars and light trucks. But technology like new means of propulsion (electric motors) and stronger, lightweight steel were still a few years off. So to meet the rising fuel-efficiency requirements, companies continued to removed pounds and reduced the track and wheelbases of their vehicles, making cars smaller and lighter. Would this cause the predicted safety crisis?

MORE RESEARCH

Dr. Charles Kahane was well qualified to answer the safety question. A lifelong analyst at NHTSA, he had used federal crash data in his research for decades. His work included trend analysis of the fatal crash statistics, discussed in Chapter 4, and the problem of vehicle incompatibility, discussed in Chapter 6.

In 1995, Kahane submitted a draft study on the tie between vehicle weight and fatality risk to the National Research Council's Transportation Research Board (TRB). The purpose was to gain feedback on his research and conclusions. Kahane used fatal crash data and statistical modeling methods to conclude that reducing the weight of cars would cause more deaths.

The TRB committee did not endorse these results. In their opinion, Kahane had not gone far enough to account for driver behavior in a crash. They suggested that aggressive drivers tended to drive smaller cars that were sportier and more powerful. Smaller cars were not less safe – it was the driver.

Kahane dug deeper. He used data such as prior license suspensions and speeding violations to measure aggression. He found these behaviors occurred in average- or higher-weight vehicles, not small cars. Furthermore, aggressive driving was related to more horsepower and high performance, not necessarily vehicle size. When he repeated his analysis excluding these aggressive drivers, the results did not change.

In his final 1997 report, Kahane wrote that there were:

- **Three hundred more deaths when manufacturers lightened cars by 100 pounds**, while light trucks and other vehicles remained unchanged, and

- **Forty fewer fatalities when manufacturers reduced the weight of light trucks by 100 pounds**, while cars and other vehicles remained unchanged.

The NHTSA analyst was identifying a trend. It did not appear to be good practice to lighten cars, but it might be beneficial to lighten pickup trucks,

SUVs, and minivans. In a nutshell, lightening these bigger vehicles might make them less harmful.

Building on his vast knowledge of the factors that contribute to fatal crashes, Kahane included multiple factors in his 1997 analysis. Of particular concern was the driver's age and gender. He recognized:

> Size-safety analyses must control for age and gender, otherwise, they will attribute safety problems to small vehicles that are actually due to the young drivers who use those vehicles.

For instance,

- Younger drivers tend to drive smaller, lighter vehicles and are less experienced.

- Younger men tend to drive fast and take more chances than other drivers. Younger women are more vulnerable than young men in a crash.

- Women drive more cautiously than men.

- Older drivers tend to drive bigger cars, drive with more caution, and are more vulnerable to injury and death in a crash.

- Older men have higher fatalities rates than older women.

When Kahane controlled for the driver's age and gender in his statistical models, he essentially was building separate models for drivers of each of the age-gender combinations. In this way, each model captured the effect of removing 100 pounds from a vehicle. The final results of the analyses, highlighted above, came from bringing together all the results from all the models.

However, there was an important caveat to these analyses. Kahane could not directly study the effect of removing pounds from vehicles. After all, he was working with historical fatal crash information. People perished in vehicles. The vehicles were small or large; weighed 2,500 or 4,000 pounds. There was no way to change the weight of the vehicle in the crash. Nor were there enough data to examine specific make-models before and after manufacturers removed weight to improve fuel economy.

Instead, Kahane used a **cross-sectional analysis**. He compared the fatality rates of vehicles weighing a specific amount with rates of similar vehicles weighing 100 pounds less across make-models. Thus, the comparison vehicles and their drivers might have been different from the base models under scrutiny. For instance, if the fatality rate of a heavier Ford vehicle is better (smaller) than the fatality rate of a lighter Chevrolet vehicle, there is no way to be certain that the better Ford rate is due to being heavier. The Ford vehicle might have a better structural design or better seatbelts. Nonetheless, this comparison among vehicles of various weights acted as a surrogate about what might happen if vehicles shed pounds.

Kahane also assumed that the driver and environment of the comparison vehicles were similar. While the modeling controlled for driver age and gender, as well as the location of the crash (urban versus rural), time of crash (daylight versus night), and type of road, he knew there could be other factors that would contribute to different crash outcomes. For example, perhaps Ford drivers were more cautious than Chevrolet drivers. It was not possible for Kahane to control for such unknowns in this analysis.

In addition, lightening vehicles generally resulted in vehicles getting smaller. In his analysis, Kahane could not separate the effect of reducing weight from reducing size. So, when he drew conclusions about the impact of weight reduction, Kahane unavoidably studied the impact of making vehicles smaller, as well. As I will discuss below, this became part of the debate about the statistical analyses.

In spite of its limitations, Kahane's study in 1997 was the most complete research to date. Still, it did not satisfy everyone.

MORE CONTROVERSY

In 2000, Congress requested the National Academy of Sciences (NAS) work with the federal Department of Transportation to study the impact of the CAFE standards. The congressional directives covered a wide range of issues including assessing the impact of the standards on:

- the U.S. economy, in general;
- the automotive industry, including vehicle design and consumer choices, in particular; and
- vehicle safety.

Congress wanted the NAS to evaluate how well the standard was working and if there were aspects of the standard that needed fixing in future regulations. The committee understood that the issues confronting them were intertwined when they wrote their 2002 report:

> For example, if fuel economy standards were raised, the manner in which automotive manufacturers respond would affect the purchase price, attributes, and performance of their vehicles. For this reason, the mix of vehicles that a given manufacturer sells could change, perhaps resulting in a greater proportion of smaller and lighter vehicles; this, in turn, could have safety implications, depending on the eventual mix of vehicles that ended up on the road. If consumers are not satisfied with the more fuel-efficient vehicles, that in turn could affect vehicle sales, profits, and employment in the industry. Future effects would also depend greatly on the real price of gasoline; if it is low, consumers would have little interest in fuel-efficient vehicles. High fuel prices would have just the opposite effect. In addition, depending on the level at which fuel economy targets are set and the time the companies have to implement changes, differential impacts across manufacturers would probably occur depending on the types of vehicles they sell and their competitive position in the marketplace.

At the completion of their work, the 13-member committee of the NAS made 15 recommendations. There was agreement on all the suggestions except one: safety.

All but two of the committee members reached a conclusion similar to that of Crandall and Graham in 1989. This majority believed that car occupants died because vehicles became lighter to meet the CAFE standards. However, they remained cautious, as seen in the following statement from their 2002 report. (Italics added for emphasis.)

> In summary, the majority of the committee finds that the downsizing and weight reduction that occurred in the late 1970s and early 1980s *most likely produced* between 1,300 and 2,600 crash fatalities and between 13,000 and 256,000 serious injuries in 1993. The proportion of these casualties attributable to CAFE standards is *uncertain*. It is not clear that significant weight reduction can be achieved in the future without some downsizing, and similar downsizing *would be expected* to produce similar results. Even if weight reduction occurred without downsizing, casualties *would*

be expected to increase. Thus, any increase in CAFE as currently structured *could produce* additional road casualties, unless it is specifically targeted at the largest, heaviest light trucks.

Overall, the committee members acknowledged that much progress had been made to improve road safety. Among those trends were the decrease in alcohol-impaired driving, more seat belt use, and better road construction. Such improvements made it difficult to identify and isolate a safety tax associated with the CAFE standards. However, they pointed to the change in vehicles weights. As shown in Figure 7.4, cars had lost 700 pounds and light trucks 300 pounds between 1976 and 1993. They concluded that if manufacturers continued to down weight and/or downsize cars to meet fuel efficiency standards, then more lives would be lost.

But two members of the committee were not convinced. Dr. David Green of Oakridge National Laboratory and Dr. Maryann Keller, retired from priceline.com, wrote an appendix to the final committee report in protest. They emphasized that there were winners and losers in a collision due to the disparity in vehicle weight. Therefore, the emphasis should be on the relative weight of vehicles, not weight reduction of some

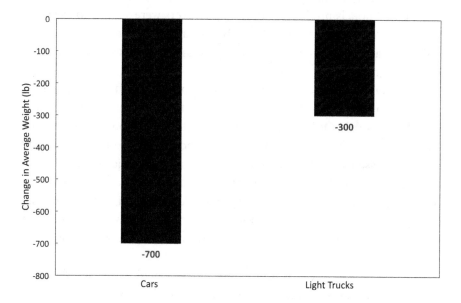

FIGURE 7.4 Change in the average weight (lb) of passenger cars and light trucks between 1976 and 1993.

vehicles. A better formula for safety was to push for all vehicles to be similar weight. In addition, there was a lack of information and research into the role of the driver. The dissenters wrote in the appendix of the 2002 report:

> Part of the difficulty of estimating the true relationships between vehicle weight and highway safety is empirical: Reality presents us with poorly designed experiments and incomplete data. For example, driver age is linearly related to vehicle weight [younger drivers tend to drive smaller vehicles], and vehicle weight, size, and engine power are all strongly correlated. This makes it difficult to disentangle driver effects from vehicle effects.

While they agreed that Kahane's 1997 study represented the best information to date, in their view, it still fell short.

A MAJOR CHANGE TO THE STANDARD

Though the NAS committee could not reach consensus about the safety impact of the CAFE standards, the members agreed that the standards should be overhauled to discourage the production of smaller, lighter vehicles. Furthermore, the system at that time was insensitive to the diversity among the models produced. All manufacturers were subject to the same fuel economy target as shown in Table 7.1 and illustrated in Figure 7.2. This standard did not allow for different manufacturers serving different market segments.

For instance, a manufacturer that sold two models of luxury passenger cars had to meet the same standard as one that sold ten different, more affordable models. Or if a maker had a market niche of sports vehicles, that company had to meet the same fuel efficiency target as one that focused on large passenger cars. Regardless of their mix of vehicles, each manufacturer had to conform to a single standard or face a financial penalty. As described earlier, this type of the standard led to manufacturers balancing larger vehicles with lighter and smaller vehicles. For example, some experts believe that Mercedes-Benz bought the mini SmartCar for just this reason.

The NAS committee recommended an attribute-based standard. In other words, pick something about vehicles, such as weight, wheelbase, or track width, and set the fuel economy standard based on that attribute.

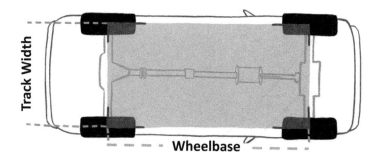

FIGURE 7.5   The product of the wheelbase and the track width equals the vehicle footprint.

In addition, the new standard should not incentivize manufacturers to make smaller and lighter vehicles. Rather, each vehicle produced by the company, depending on the attribute, should have a target fuel economy. Each company would then have its own standard depending on its mix of vehicles sold.

Congress passed the Energy Independence and Securities Act in December 2007. That act mandated that the CAFE standards be attribute-based. After much discussion, NHTSA chose the footprint of the vehicle as the attribute. As shown in Figure 7.5, the footprint is approximately the area of the rectangle formed by the touchpoints of the tires. That is, the wheelbase multiplied by the track width. Unlike weight, footprint is not dependent on components of the vehicle, such as engine size. Also, manufacturers could, for example, increase vehicle overhang in front of the tires to increase crush space for safety or reduce weight, but not necessarily size, to meet a target value.

The fuel efficiency targets for cars became a series of line graphs, similar to those shown in Figure 7.6 (top) for 2021 and 2026. An important feature of these graphs is that smaller vehicles had to meet higher mpg targets. The downward sloping line in Figure 7.6 (top) shows that as a vehicle's footprint gets bigger, the target mpg gradually gets lower. The higher mpg for the smaller footprint is a disincentive to reduce vehicle size. Lower mpg targets incentivized larger footprints vehicles that are typically safer.

Similar curves were set for light trucks, as shown in Figure 7.6 (bottom). These included any vehicle built on a truck frame, such as pickups, SUVs, and minivans.

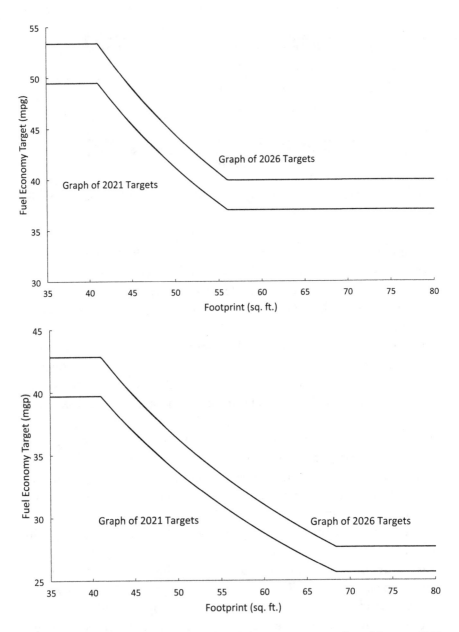

FIGURE 7.6 (top) Fuel economy targets for passenger cars of model years 2021 and 2026. (bottom) Fuel economy targets for light trucks of model years 2021 and 2026.

NHTSA developed these graphs using statistical analyses linking fuel consumption and footprint, together with information – sometimes confidential – supplied by manufacturers. Market forces, consumer behavior, and fuel prices were among the factors taken into account. The parallel line graphs in Figure 7.6 show that over time, the target values increased. In other words, the manufacturers had to reach for higher and higher fuel efficiency targets or face penalties. However, pickups, SUVs, and minivans always have less fuel efficient target values than cars with the same footprint.

Under this revised standard, each manufacturer's vehicle model would have its own CAFE target because each model would have a different target value depending on its footprint, as illustrated in Figure 7.7a. Here again, the stacked vehicles represent sales volume, but now each vehicle has its own target mpg standard. Then based on sales volume and vehicle mix, each manufacturer would have its own composite company standard target, shown as the larger target in Figure 7.7a. If some of the models did not meet their individual target, then it was still possible to use other models to balance the production/sales with other vehicles that performed better than their target. Companies had more flexibility though credits and penalties would remain part of the standard, as shown in Figure 7.7b.

For example,

- In 2021, the Accord's footprint was about 48.7 square feet (with a wheelbase of about 9.3 feet and a track width of about 5.25 feet) giving it a target of 42.2 mpg.

- Honda makes various versions of the 2021 Accord with fuel economies between 26 and 33 mpg. These clearly miss their target of 42.2 mpg.

- However, there is also the 2021 Accord Hybrid models getting between 43 and 48 mpg. The non-hybrid vehicles with missed targets could be balanced by the hybrid model.

As intended, the new standard encouraged new technologies with more fuel efficiencies.

Statisticians tasked with assessing the safety implications of this new standard faced huge challenges. Researchers had to perform a different analysis than the one used for the 1997 report on the impact of reducing vehicle weight. That is, they had to study the impact of reducing vehicle weight, while maintaining (not changing) a vehicle's footprint.

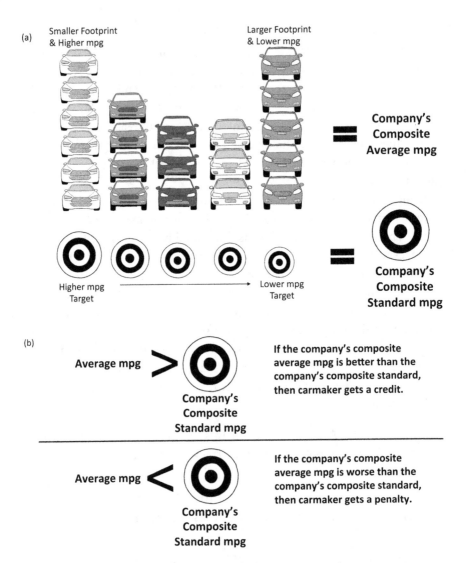

FIGURE 7.7 (a) Illustration of new footprint-based, company CAFE standard and composite average. (b) Illustration of the credits and penalties of new footprint-based, company CAFE standard.

GETTING MORE COMPLICATED

The 2010s was a very busy decade for predicting the safety impact of the CAFE standards. NHTSA analysts produced five major statistical reports. With each new report, the analysis was more intricate. In addition, NHTSA sponsored two workshops to discuss and get feedback on their methodologies and predictions.

During this decade, there were more stories about the harmful effects of fossil fuel. Concern for a warming planet grew. And not just in Washington and the halls of academia.

Kurt Martin remembers walking on ice around his commercial fishing boat at Christmas time in the 1990s. Back then, the water temperature in Cape Cod, Massachusetts only reached above 60 degrees a few times a year – at the peak of summer. Now the water sometimes jumps to the low 70s, and there has not been thick winter ice for years.

For nearly three decades, Martin kept a logbook detailing his catch amounts, species abundance, weather conditions, water temperatures, and more. It's clear to him why local fishermen can no longer catch the cape's namesake and scientists confirm it.

The waters off of the New England coast are warming faster than 99% of the world's oceans, according to NASA's Earth Observatory. Warming air temperatures, rising greenhouse gas concentrations, and melting ice in Greenland and the Arctic Ocean have altered ocean circulation patterns in the region.

Cape Cod fishermen now bring in dogfish, skate, and sea bass. It's rare to catch any cod off Cape Cod.

Carmakers promoted fuel-efficient vehicles, especially electric cars, as the solution. That is, new technology would save the day. However, there were many unknowns. Would consumers accept the new technology? How fast would these vehicles sell? Were they safer than combustion-powered vehicles?

NHTSA analysts had to use the only real data available for their safety predictions. That is, they used historical data to forecast years into the future. Figure 7.8 illustrates an example. For the 2020 report, the analysis used data about vehicles from model years 2004 through 2011 in fatal crashes during the years 2006 through 2012. The results were used to forecast the safety impact of the CAFE standards for vehicles model years 2022 through 2025. A major reason for this severe lag in the data was the expense of collecting and preparing the data for the analysis.

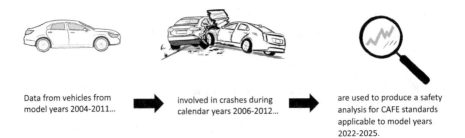

FIGURE 7.8   Illustration of the data used to generate NHTSA's safety impact of the 2020 CAFE standard.

To fine tune their results, the NHTSA analysts created more complicated statistical models. For instance, different classes of vehicles were analyzed separately. By 2020, there were five classes:

- Lighter cars
- Heavier cars
- Crossovers/minivans
- Lighter light trucks
- Heavier light trucks

It took work to separate the lighter from the heavier vehicles. NHTSA analysts used the median weight (i.e., the 50th percentile) to separate the lighter and heavier vehicles. As shown in Figure 7.9, this value has been increasing over time. For instance, in the 2010 report, the median weights were 2,950 pounds for cars and 3,870 for light trucks. By 2020, those medians were 3,201 and 5,014. Cars had gained weight, but not as much as light trucks.

In addition to building separate forecasting models for different groups of vehicles, NHTSA researchers divided the analyses by nine different types of crashes. For instance, single-vehicle rollover crashes were distinguished from vehicle-pedestrian crashes. Car-light truck, two-vehicle crashes were separated from car-car crashes. Together, the vehicle classes and crash types made the data analysis more complex. However, it also gave analysts ample flexibility in their investigations.

In these later analyses, NHTSA investigators wanted to see if there was a so-called societal benefit for lightening vehicles. Whereas most car safety studies focus on occupant fatalities and occupant risk, these CAFE analyses took a broader view. Instead of *only* counting fatalities in vehicles, these

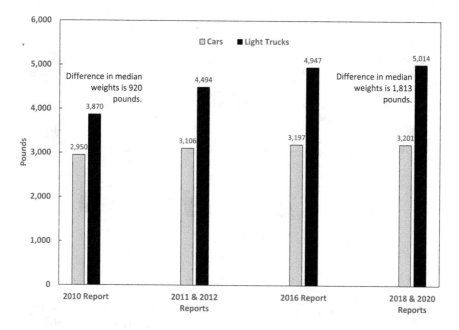

FIGURE 7.9 The median weights of vehicles used in NHTSA's safety impact analyses.

investigations counted fatalities *inside and outside the vehicle*, including cyclists, pedestrians, and motorcyclists. For example, if lightening vehicles did make them more maneuverable, without sacrificing occupant safety, then maybe fewer cyclists would be killed.

Each of NHTSA's five reports and detailed analyses assumed that the CAFE standards – with their increasing fuel efficiencies – would be met by reducing the weight of the vehicles, but without changing the footprint. However, that assumption and various aspects of the statistical models did not sway all interested parties.

THE BIG DEBATE

NHTSA established multiple public dockets in the 2010s. Organizations submitted hundreds of commentaries about statistical and engineering issues associated with NHTSA's reports. These organizations provided important views and ideas for arriving at a comprehensive final safety analysis.

(a)

UNIVERSITY

NHTSA overestimates
safety costs by using
the wrong data.

Other statistical models
show no safety penalty.

Other statistical models
are less credible.

Statistical modeling problems are
due to the relationship of curb
weight, track width, and
wheelbase.

CIVIC GROUP

Using the wrong data
biases the results.

The vehicles are not
grouped correctly.
Group them by
footprint, not types.

Model is adequate, but
too simple.

GOVERNMENT

Small changes in data
yield different results—
statistical models are
not stable.

Driver responses to
vehicle changes need to
be captured in the data.

Using different data do
not produce better
statistical models.

IIHS

CALIFORNIA

CLEAN AIR
GROUP

NATIONAL
ACADEMY OF
SCIENCES

INDUSTRY

(b)

UNIVERSITY

In the future, there are likely to
be more crashes involving light
trucks, but overall fewer crashes
due to driver-assist technology.

Vehicle structural and safety
equipment influence occupant
injuries in heavier vehicles.

New designs make it possible to
reduce vehicle weight without
change in size.

CIVIC GROUP

Same-size vehicles could have
different weights due to design or
horsepower.

Vehicle size and weight effect
occupant injury differently.

GOVERNMENT

Emphasize future engineering and
vehicle design.

Provide insight into how much to
increase footprint to offset
weight reduction.

Reduce the weight of vehicles, but
make them more crashworthy.

IIHS

CALIFORNIA

CLEAN AIR
GROUP

NATIONAL
ACADEMY OF
SCIENCES

INDUSTRY

FIGURE 7.10   (a) Sample feedback on data and statistical issues in NHTSA's analyses. (b) Sample feedback on engineering issues in NHTSA's analyses.

Many comments from key organizations took aim at data and statistical issues as represented in Figure 7.10a. For instance, some claimed that NHTSA overestimated safety costs by using the wrong data. Others suggested statistical modeling problems occurred because of the relationship among vehicle curb weight, track width, and wheelbase. This made the analysis sensitive to small changes in the data. Some considered NHTSA's model as adequate, but too simple. Some critics doubted that the data captured driver behavior when vehicle weight changed.

Engineering issues also emerged as potential problems with NHTSA's analysis as shown in Figure 7.10b. These issues tended to center around the likely changes in future vehicles. For example, some suggested that historical data did not reflect the higher number of light trucks and vehicles with driver-assist technologies. Plus, newer designs and high strength/low-weight materials would improve occupant safety. Some believed that the emphasis should be on reducing the weight of vehicles while pushing the industry to adopt these newer materials and designs to improve vehicle crashworthiness.

Again, every organization voiced an opinion about these types of engineering issues. However, groups advocating for cleaner air and energy conservation, such as the State of California, tended to predict that future vehicles would be safer and more fuel efficient. Industry representatives tended to share this optimistic outlook, but they also voiced caution about the added cost of adopting innovation, the role of the driver, and the challenge of meeting changing market demand.

NHTSA analysts had a huge task to respond to this feedback. Some responses required additional analysis. Other suggestions NHTSA found to be technically infeasible, i.e., the data were not available or resulted in too small sample sizes for reliable results. After processing all the inputs, NHTSA either altered its original analysis or stuck to its previous results.

In their 2020 Final Rule on the CAFE standards applicable for vehicles in model years 2021–2026, the NHTSA analysts came to a new conclusion. There would be no significant change in fatality risk due to meeting the CAFE standards. According to their analysis, the CAFE safety tax suggested in the 1980s no longer existed.

NOTE FROM THE AUTHOR

They began as a response to the Arab oil embargo nearly a half century ago. Now in many people's minds, the CAFE standards are part of the movement to slow climate change. The trend toward more fuel-efficient vehicles is accelerating.

Every major manufacturer has plans to move away from carbon-based fuel vehicles. Plus, investors are sinking billions of dollars into new startups making electric and hydrogen-fueled vehicles. Car buyers soon will have more choices. Even so, there are nearly 280 million registered vehicles and only about 5% of the 13–14 million vehicles sold annually currently are these alternative powered vehicles. It will be decades before the nation understands the safety effects of these new types of vehicles.

There is another trend that could have an even greater impact on fuel economy and safety forecasts. Car owners are retaining their vehicles longer. In 1995, the average age of automobiles and light trucks was 8.4 years. In 2019, estimates put that number at 11.8 years. And keeping a pickup truck more than 15 years is fairly common.

President Trump used this trend to argue against increasing the fuel efficiency targets. The reasoning focused on the added costs to carmakers for redesigning vehicles to meet the rising standards. Added costs result in higher prices of new vehicles. That would spur more consumers to retain their older vehicles. These older vehicles do not have the newest safety technologies, such as forward collision warning, lane departure warning, or pedestrian detection. The net result, argued the Trump administration, would be more deaths on the roadways.

As this manuscript was going to press, President Biden was working to reverse the Trump-era standards and increase fuel efficiencies.

In my opinion, the CAFE debate illustrates the best and most extensive data-intensive investigation in the history of car safety. It is also a perfect example of the challenges facing statisticians working on federal policies. Programs change with administrations. New industries arise. International events disrupt plans and fuel prices. Human interest stories, such as the Cape Code fisherman, convince people to make different choices. And yet, statisticians are tasked with making predictions 5–10 years into the future. The challenge is huge. Let's hope that talented statisticians in our nation continue to rise to the occasion.

NOTE

1. National Research Council. (2002). *Effectiveness and impact of Corporate Average Fuel Economy (CAFE) standards.* The National Academies Press, p. 11. https://doi.org/10.17226/10172.

CHAPTER 8

The Safety Ratings Debate

The auto industry was infuriated that the government was inform-
ing consumers about the actual crash performance of its vehicles by
make and model.

*—JOAN CLAYBROOK & ADVOCATES FOR
HIGHWAY AND AUTO SAFETY (2019)*[1]

President Carter must have known that he was poking the bear. Chief automobile regulators usually came from *within* the industry. After all, these insiders
were familiar with the issues, and they understood how to get things done.
But Carter's nomination of attorney Joan Claybrook to head the National
Highway Traffic Safety Administration (NHTSA) came out of left field
because she had led Ralph Nader's advocacy group, Citizen Watch. Just the
mention of her name made heads spin in the boardrooms of carmakers.

However, Claybrook had other qualifications. As a congressional aide,
she helped to pass the Transportation Act of 1966. In addition, she had
served as special assistant to the first NHTSA director, William Haddon. In
that role, she shared the credit for shaping the department and for establish-
ing the first, federal auto safety standards. While her confirmation hearing
in 1977 was contentious, Claybrook ultimately won the appointment.

Claybrook's tenure at NHTSA was short, ending with Carter's departure from office in January 1981. However, her impact is among the most
enduring of all leaders in car safety. This is due, in part, to her role as a key
architect of NHTSA's consumer testing program, the New Car Assessment

DOI: 10.1201/9781003035343-8

Program (NCAP). This program would not only save countless lives in the United States, but also inspire countries worldwide to start their own safety testing programs.

The NCAP was the world's first public program that specifically tested vehicles for crashworthiness. Before the NCAP, car buyers had to rely on industry marketing or proprietary testing to decide if a vehicle was safe. With the NCAP, consumers could gain access to information about a car's performance in laboratory crash tests, which were conducted at relatively high speed. NHTSA designed the tests to gauge how well a vehicle protected its driver in an accident. In the late 1970s, this was a revolutionary step toward arming car buyers with objective and potentially life-saving information. Then in 1993, NHTSA took another important step by making NCAP data clearer to the general public by introducing the 5-star safety rating system.

Now, the safety performance of a vehicle is easy to interpret. At the high-end, a 5-star rating means that the car demonstrated the highest occupant protection in the tests. At the low-end, a 1-star rating implies the poorest protection. The purpose of the 5-star program is twofold:

1. Raise consumer awareness and prompt them to buy safer cars, and

2. Encourage carmakers to earn more stars by building safer cars.

If more consumers become aware of the different safety ratings of cars, then this prompts manufacturers to build safer cars. The intent is to promote the notion that safety sells.

Today, most buyers check the window sticker on a new car at the dealership to find the listed price of the vehicle. But by law, this sticker also must display the NHTSA safety rating. Two so-called Monroney labels are illustrated in Figure 8.1 (top) for the 2020 Mazda Model 3 (MM3), a small sedan, and Figure 8.1 (bottom) for the 2020 Ram 1500 Classic, a large pickup with a regular cab and 4-wheel drive.

Generally, car buyers assume that they will be safer in vehicles with more stars. However, there is an important caveat in the fine print, circled in Figure 8.1 (top and bottom). A vehicle's star safety rating "Should ONLY be compared to other vehicles of similar size and weight." In other words, you cannot and should not assume that the 5-star rated MM3 is safer than the 4-star rated Ram 1500. That is, if two vehicles are not close in size and weight, then their ratings cannot be compared.

GOVERNMENT 5-STAR SAFETY RATINGS

Overall Vehicle Score ★★★★★
Based on the combined ratings of frontal, side and rollover.   [Emphasis added]
Should ONLY be compared to other vehicles of similar size and weight.

Frontal	**Driver**	★★★★★
Crash	**Passenger**	★★★★★

Based on the risk of injury in a frontal impact.   [Emphasis added]
Should ONLY be compared to other vehicles of similar size and weight.

Side	**Front Seat**	★★★★★
Crash	**Rear Seat**	★★★★★

Based on the risk of injury in a side impact.

Rollover ★★★★★
Based on the risk of rollover in a single-vehicle crash.

Star ratings range from 1 to 5 stars (★★★★★) with 5 being the highest
Source: National Highway Traffic Safety Administration (NHTSA)
www.safercar.gov or 1-888-327-4236

GOVERNMENT 5-STAR SAFETY RATINGS

Overall Vehicle Score ★★★★
Based on the combined ratings of frontal, side and rollover.   [Emphasis added]
Should ONLY be compared to other vehicles of similar size and weight.

Frontal	**Driver**	★★★★
Crash	**Passenger**	★★★★

Based on the risk of injury in a frontal impact.   [Emphasis added]
Should ONLY be compared to other vehicles of similar size and weight.

Side	**Front Seat**	★★★★★
Crash	**Rear Seat**	★★★★★

Based on the risk of injury in a side impact.

Rollover ★★★
Based on the risk of rollover in a single-vehicle crash.

Star ratings range from 1 to 5 stars (★★★★★) with 5 being the highest
Source: National Highway Traffic Safety Administration (NHTSA)
www.safercar.gov or 1-888-327-4236

FIGURE 8.1   (top) The safety section of the new car Monroney label for the 2020 Mazda Model 3 with emphasis added. (bottom) The safety section of the new car Monroney label for the 2020 Ram 1500 Classic regular cab 4-wheel drive pickup with emphasis added.

In order to understand this limitation, you first need to grasp how a vehicle earns a certain rating.

LABORATORY TESTS FOR SAFETY RATINGS

The 5-star safety ratings stem from a series of laboratory tests. Figure 8.2 shows the general steps that NHTSA uses to assign ratings. First, they strap dummies made of vinyl, rubber, and steel into a brand-new car. They're meant to represent you and your family. Then, they crash the car. During and after the crash, engineers measure the stresses and impacts on the dummies. For example, how much did the dummy's neck extend or the chest compress during the crash?

The engineers then compare these measurements to injury risk curves. The curves estimate the chances of a human occupant being severely injured. The engineers use a different injury curve depending on where the stress occurred on the dummy. They focus on those parts of the body that can lead to serious injury or death. For example, severe extensions of a person's neck can lead to paralysis; deep depressions in the chest can be fatal.

These injury curves date back decades, and they draw from data created by using human volunteers, cadavers, and prototype dummies. Most importantly, the injury curves typically are calibrated to a 35-year-old male body.

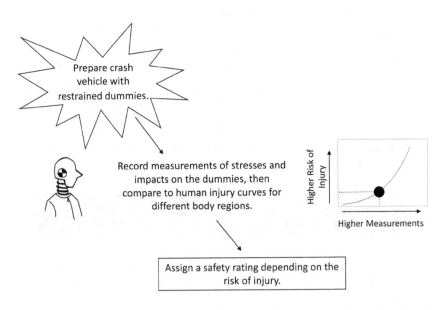

FIGURE 8.2 Illustration of the general process of assigning safety ratings based on laboratory tests.

Some NCAP tests simply scale down the average male dummy to represent a small female. When the engineers use this smaller dummy, as you will see in the tests described below, then the male injury curves also are scaled down to depict female injury risks. The practice of scaling to represent females has become a hotly debated topic.

Periodically, NHTSA proposes changes to the NCAP. The proposal is posted in the Federal Register, a docket is opened, interested parties provide feedback and a debate ensues. In this chapter, I start by describing the NCAP tests that have existed for over a decade and then describe some of the current debatable issues in these tests.

DEEPER DIVE INTO THE NCAP TESTS

The first NCAP test is a frontal one that consists of crashing a vehicle into a rigid barrier at 35 mph, as illustrated in Figure 8.3a. This test intends to mimic a frontal collision with another car. The dummies used in the test are the mid-size male (172 lb and 5'9" tall) in the driver's seat and a small female (109.7 lb and 4'11" tall) in the right-front passenger seat.

The crash resembles a vehicle colliding head-on into another vehicle that is about the same size and weight. For example, as shown in Figure 8.1 (top), the small 2020 MM3 received 5 stars for both the driver and passenger in the frontal-crash test. In other words, the dummies fared very well in a simulated, head-on crash with another small sedan that was similar in weight to the MM3.

On the other hand, the Ram pickup only got 4 stars on its frontal test (see Figure 8.1 (bottom)). In that test, the front-seated dummies were stand-ins for two people in a Ram pickup having a head-on collision with another large pickup that was similar in weight to the Ram. The Ram received a lower star rating than the MM3, but the tests results cannot be compared.

What would happen if an MM3 had a frontal collision with a Ram 1500? Which occupants would be better protected? Chapter 6 discussion on incompatibility addressed this vital question. The occupants of the heavier vehicle are typically better protected than occupants of the lighter vehicle. This fact is based on the physics of the crash, as well as fatal crash statistics. As you can see, the star rating system cannot be applied to a frontal collision between a large and small vehicle.

How about the other tests in NCAP's 5-star safety rating system?

The two NCAP laboratory tests that measure side-impact crashes tell a different story. The results of the MM3 and Ram 1500 can be directly compared. That is because vehicles experience the same forces, regardless of the vehicle.

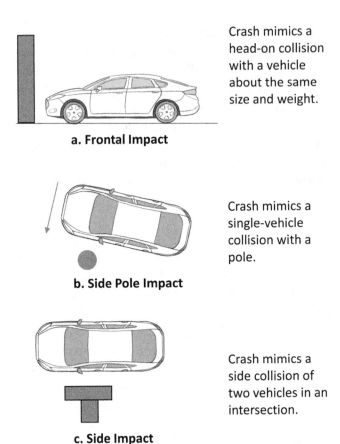

Crash mimics a head-on collision with a vehicle about the same size and weight.

a. Frontal Impact

Crash mimics a single-vehicle collision with a pole.

b. Side Pole Impact

Crash mimics a side collision of two vehicles in an intersection.

c. Side Impact

FIGURE 8.3 (a) Frontal impact test designed to mimic a head-on collision with a vehicle about the same size and weight. (b) Side pole impact test designed to mimic a single-vehicle collision with a pole. (c) Side impact test designed to mimic a side collision of two vehicles in an intersection.

In the pole test, as illustrated in Figure 8.3b, a vehicle moving at 20 mph strikes a pole on the driver's side at a 75-degree angle. A dummy representing a small woman is in the driver seat. The pole depicts, for example, a telephone pole. This type of crash comprises about 20% of severe side-impacts, according to historical fatality and injury data.

The other 80% of severe side-impact crashes are represented by a broadside test, as illustrated in Figure 8.3c. In this test, an average-size, male dummy sits in the driver's seat. A dummy representing a small female sits in the rear seat behind the driver. The test launches a *moving deformable object* that weighs 3,015 lb into the side of the test vehicle at 38.5 mph. The object mimics a vehicle, similar to a 2018 Honda Civic, broadsiding the test vehicle at an intersection.

The results of the two, side-impact tests are combined into one, side-crash rating. As shown in Figure 8.1, both the MM3 and Ram 1500 got top ratings, 5 stars. In other words, the MM3 protected its occupant as well as the Ram 1500, according to these tests.

The NCAP also includes a rollover test. It is designed to assess the rollover risk in a single-vehicle crash, and it does not use dummies with injury curves. The rating is based largely on a vehicle's static stability factor (SSF). Illustrated in Figure 8.4, this factor is strongly related to how close the vehicle sits to the ground, as represented by the location of the center of gravity. It is computed as:

$$Static\ Stability\ Factor\ (SSF) = \frac{\frac{1}{2}\ Track\ Width}{Height\ of\ the\ Center\ of\ Gravity}$$

A vehicle with a smaller SSF value has a higher rollover risk, while a vehicle with a larger SSF has a lower risk because it is more stable.

In the rollover test, the vehicle performs a set of maneuvers. The purpose is to try to initiate a roll. If a vehicle's wheels lift off the pavement, then this result points to a lower safety rating. Tire lift usually occurs for vehicles with low SSF. As illustrated in Figure 8.4, sedans usually sit closer to the ground, where light trucks sit higher up. Consistent with this, the MM3 small sedan has a relatively high SSF of 1.45. It received 5 stars in the rollover test. The Ram 1500 has a relatively low SSF of 1.15. It only received 3 stars in the rollover test. In other words, the Ram has a higher risk of a rollover crash.

Larger SSF implies lower rollover risk.

Smaller SSF implies higher rollover risk.

FIGURE 8.4   Vehicles with different static stability factor (SSF) values.

FIGURE 8.5   Illustration of how stars are assigned in the federal 5-star safety rating system based on the vehicle relative risk.

The overall vehicle star rating combines the results of all the tests into a composite score called the vehicle relative risk (VRR). In combining the results, the system takes into account the importance of each type of crash, based on historical fatality patterns. That is, how people die in crashes. The results of the frontal test are weighted the highest in arriving at the VVR at 42%. The side test results are weighted at 33% and the rollover at 25%.

The value of the VRR determines the number of stars of the vehicle, as shown in Figure 8.5. If the VRR is less than 10%, then the vehicle receives a 5-star safety rating. A relative risk of 10%-15% results in a 4-star rating, and so forth.

In summary, the overall vehicle star rating – as illustrated on the Monroney labels in Figure 8.1 – gives the consumer information about test results. While the tests aim to simulate actual crashes, they are limited by the procedures designed into the tests. As such, the rating carries the important warning: "Should ONLY be compared to other vehicles of similar size and weight."

But sometimes, a manufacturer or salesman will ignore this limitation of the 5-star safety rating system.

In October 2018, Tesla promoted its ratings for the Model 3 on Twitter saying, "There is no safer car in the world than a Tesla." The sedan had achieved the best score of any vehicle tested in the 5-star safety rating system. The post pointed to the company's October 7 blog that states:

> Many companies try to build cars that perform well in crash tests, and every car company claims their vehicles are safe. But when a crash happens in real life, these test results show that if you are driving a Tesla, you have the best chance of avoiding serious injury.

In response, a NHTSA attorney wrote a letter to Tesla emphasizing the fact that these tests results should not be used to make such far-reaching global statements. In particular, the letter stated:

> The nature of the (frontal-crash) test makes it impossible to compare results of vehicles that vary in weight by more than 250 pounds.

In other words, just as the ratings of the Mazda could not be compared to the Ram, neither could the Tesla Model 3 results be compared to other heavier vehicles. Furthermore, it is "impossible" to say, based on this test, what would happen if a Model 3 were in a head-on collision with a Ram 1500 truck.

Most experts consider the 5-star safety rating system a successful consumer and safety-promoting program. Yet on the 40th anniversary of the NCAP, this program that gave birth to the star safety rating system faced a critical report from its key architect and a group of consumer advocates.

Joan Claybrook and Advocates for Highway and Auto Safety published *NCAP at 40: Time to Return to Excellence* in October 2019. Recalling industry resistance to the NCAP, the report chastised NHTSA for losing its leadership role in the world. The U.S. NCAP was the first such program in the world, but now it lags far behind. For example, the European safety testing program, Euro NCAP, has 21 tests. These include rear-impact whiplash, child occupant protection, pedestrian and cyclist protection, and testing of new driver-assist technologies. Also, the Euro NCAP undergoes regular updates, including in 2020 when European regulators added a far-side-seated-occupant test for side impacts.

As this book is going to print, NHTSA is soliciting comments in an open docket to revise the NCAP. However, that is just one step in the process. In order to complete this revision, NHTSA analysts will need to defend or change some key features of their tests, starting with the dummies.

THE DUMMIES DEBATE

Dummies are meant to mimic the dimensions, weight, and movement of the human body in a crash. Originally developed for the U.S. Air Force in 1949, these stand-ins were based on the young, male body. After all, men were the only combatants at the time. That carryover into the car safety domain has a long legacy.

General Motors designed the 50th percentile Hybrid III male dummy, simply called the average male, in the early 1970s. The Hybrid III, still in use today, represented the median weight (172 lb) and height (5'9") of men at the time. Today, the median male weight is 192.6 lb. On the current graph of male weights, shown in Figure 8.6, the Hybrid III is closer to the 30th percentile. The median height of men today is about the same as the 1970s, coming in only a half-inch taller. One argument for continuing to use this Hybrid III is that its weight and height reflects the median weight and height of actual male crash victims.

The dummy used to represent a small woman in three NCAP tests is a scaled-down version of the male dummy. The intention is to represent a more fragile occupant. To this end, the dummy's measurements are based on the 5th percentile of women in the early 1970s, with a weight of 110 lb and height of 5'. Today's 5th percentile woman is only slightly lighter and shorter at 109.7 lb and 4'11". Because this dummy is a scaled-down version of the male dummy, some researchers consider it to be more like a 12-year-old boy than a mature woman.

Figure 8.7 summarizes some of the major reasons why some experts argue that a scaled-down, male body is not a suitable stand-in for the female physique. Research into the biological differences and the increased risks for women in car crashes has been underway for decades. Among the most

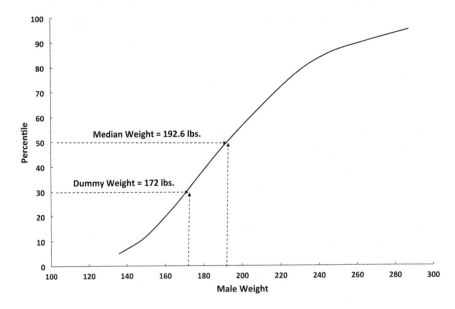

FIGURE 8.6   Percentile graph of male weight collected during years 2015–2018.

Females have 47% higher odds of severe injuries, compared to males in similar crashes.

Women have significant differences in stiffness and strength in ligaments and tendons compared to men.

Females have a greater risk of injuries to lower extremities, i.e., legs, knees, ankles, and feet.

Vertebra shapes are different in spinal column.

In rollover crashes, females are more likely to experience severe or fatal injuries than males.

Lap belts fit women and men differently due to pelvis shape and orientation.

FIGURE 8.7 Highlights of biological differences and increased risk of women in cars.

recent and noteworthy research is that sponsored by the European Union and led by Dr. Astrid Linder at the Swedish National Road and Transport Research Institute. In fact, the whiplash tests used in the Euro NCAP program are influenced by the female-centric research done by these scientists.

Women generally are more vulnerable in car crashes than men, because of differences in the structural, muscular, and vascular makeup of their bodies. For example, a 2011 study from the University of Michigan found that belted women had 47% higher odds of severe injuries than belted men in a comparable crash, when controlling for weight and body mass. Women also are more likely to be severely injured in rollover crashes.

In addition, women typically are shorter than men. When driving, they sit closer to the steering wheel to reach the pedals, putting them at greater risk of internal injuries in frontal collisions. As a result, it could be said that the driver's seat is not designed for women. This is noteworthy given that the number of female-registered drivers surpassed males in 2005, as Figure 4.3 shows in Chapter 4. Carmakers often argue that women drive less, as shown in Figure 4.7. After all, 70% of the deaths in vehicle crashes each year are males.

While there are conflicting views about who should be the focus of safety testing and design, there is evidence that the advantage of the *average-male* choice is showing up in crash statistics. One group of researchers found that males who fit the profile of the Hybrid III test dummy used in the NCAP had the fewest injuries in car crashes. Such findings lend credibility to the notion that "you get what you test to."

Chest injury risk increases sharply with age.

Aging is associated with fragility. Bone density and thickness decrease, particularly in chest and rib cage.

Severe neck and spine injuries occur twice as often in older women compared to older men.

Injury risk is 2–4 times higher in occupants 65+ compared to those aged 15-43 years.

Current chest injury curves do not reflect risks of older occupants.

FIGURE 8.8   Highlights of biological differences and increased risk of older people in cars.

Women are not the only vehicle occupants that lack representation in the current NCAP system. By 2030, people of age 65 and older are expected to make up more than 20% of the U.S. population. Figure 8.8 summarizes some of the research findings that show seniors have special needs when it comes to car crashes. For example, seniors have injury risks two to four times higher than occupants in the 15- to 43-year-old age group. Much of this is due to the fragility of older bones. Senior injury patterns are different as well.

In 2015, NHTSA proposed creating a Silver Star Safety Rating system. Like the NCAP, the idea was to spur the auto industry to focus on making cars safer for seniors. However, such a system has yet to be made public.

NHTSA has considered updating its dummies and the corresponding injury curves to better represent the growing female- and senior-driving populations. So, why have these reforms not occurred? In short, it would take a lot of data-intensive research and a strong, organizational commitment, as the following example illustrates.

In 2004, NHTSA proposed to change the average male dummy used in the side tests to one that had "enhanced, injury-assessment capabilities." A different dummy would be more sensitive to the injury patterns likely to occur in side crashes, such as spine and rib impacts as well as abdominal and pelvic injuries. The Europeans already used the proposed dummy in their testing. It had a good track record.

When NHTSA published its decision to adopt this new dummy in 2006, the final rule had 15 pages of narrative that responded to the breadth of comments received during the prior 2-year review process. These comments came from the industry, researchers, and consumer advocates. Some questioned whether the dummy was an accurate proxy for the human body. Others believed that the dummy would not record consistent measurements in the test crashes. By legal mandate, NHTSA analysts had to address all issues to defend their decision. In summary, it took a lot of NHTSA resources to make this change.

SAFER CARS OR STAR INFLATION?

A sizeable amount of this book covers the role of data in creating federal standards. The manufacturer is responsible for certifying that their vehicles conform to all applicable standards. The Office of Defects Investigation enforces this requirement. As Chapter 9 lays out, recalls often are associated with evidence that a vehicle fell short of a standard.

However, NCAP's 5-star safety rating system is not a standard or a regulation. A vehicle does not fail the NCAP. If a vehicle performs poorly in the tests, it earns 1 or 2 stars. But it still can be sold in the United States. You might think of a standard as a stick and the safety rating system as a carrot. One key purpose of the NCAP is to motivate makers to build safer vehicles. So, have cars become safer?

As Chapter 4 discussed, it is nearly impossible to give credit to any specific safety program – from seat belt mandates to drunk driving laws. However, in the case of the 5-star safety rating system, there is evidence that it has succeeded in prompting the car industry to build safer vehicles.

Manufacturers made changes, including making vehicles bigger, and improved their ratings. For example, the Nissan Versa subcompact only got 2 stars in 2011, but after a redesign it earned 5 stars in 2020. A similar pattern occurred with the Ford Escape, Toyota Corolla, and Honda Civic, which moved from 3 to 5 stars. The Subaru Forester was 4-star rated, now it's 5-star.

In 2011, 66% of the vehicles tested in NCAP received 4 stars. Only 13% received 5 stars. By 2020, those numbers had changed. Of the vehicles tested, 76% earned 5 stars and 19% earned 4 stars, as shown in Figure 8.9. In 2020, the 4- and 5-star rated vehicles represented 95% of all vehicles tested.

Some car safety experts look at the fact that 95% of vehicles receive 4 or 5 stars and declare a flaw in the system. They suggest that there is *star inflation*. Carmakers have a lot of experience with the NCAP, and they know how to build vehicles to ace the test. Maybe this is true, but even with all its flaws, has the NCAP saved lives?

NHTSA researchers compared the fatality rates of occupants in high- and low-safety rated vehicles. They found that those vehicles that were

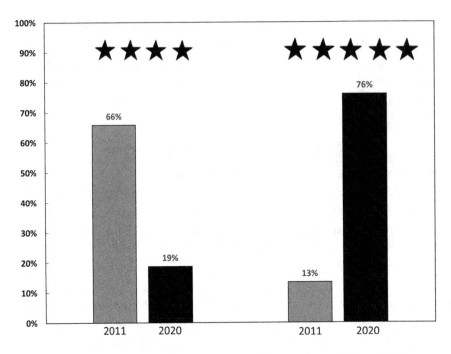

FIGURE 8.9   Percent of vehicles receiving 4 and 5 stars for model year 2011 compared to 2020.

rated safer in the frontal-crash test had 18% fewer occupant deaths than those vehicles that were rated poorly. Is this decrease due to the NCAP frontal-crash test? It seems like a logical conclusion. Also, modern vehicles are more stable and less prone to rollover risk. Can the NCAP take some credit? Maybe. But these vehicles also have electronic-stability-control systems designed to reduce rollover events.

Perhaps carmakers *have* responded to the carrot incentive. Some people in the safety community now argue that *safety sells*. In fact, there is evidence that the auto industry is using high-safety ratings as a competitive advantage.

TOP SAFETY PICKS

In addition to the 5-star safety rating system, U.S. consumers today also can check another safety rating system created by the Insurance Institute for Highway Safety (IIHS). You might be familiar with this system because manufacturers often use its *Top Safety Pick* in advertising.

Unlike NHTSA, the IIHS is free to make changes to test procedures without the public debate found in the federal docket. As a consequence, the changes to their test procedures and the rating criteria occur more frequently. Rather than provide details of the current set of tests, I discuss a couple of IIHS program features that illustrate the similarities and contrasts with the NCAP.

Like the NCAP, the IIHS researchers crash vehicles in a controlled, laboratory setting using restrained dummies in new cars. And similar to the NCAP, the frontal safety rating of different size and weight vehicles cannot be compared. For example, both the small sedan MM3 and the large pickup Ram 1500 received a Good – IIHS' top rating – in the moderate-overlap frontal test. This test resembles two similar vehicles in a frontal collision with about 40% overlap. However, again, the consumer has no information about what would happen in a similar crash between an MM3 and a Ram 1500 pickup.

In other aspects, the IIHS program complements the NCAP and raises the bar. For example, IIHS's program has two additional frontal tests. These tests mimic collisions with telephone poles or trees on both the driver and right-front passenger side. The IIHS also has a tougher side-impact test. As illustrated in Figure 8.10, the federal test uses a device that resembles a small car that weighs 3,015 lb. The IIHS test, on the other hand, uses a heavier device (currently up to 4,200 lb) that more closely resembles a light truck or SUV. The IIHS device strikes the test vehicle higher in the side panel than the NCAP device due to this difference in size.

Federal side test uses a movable barrier that resembles a car in weight and frontal structure, similar to a small car.

IIHS side test uses a movable barrier that resembles a light truck in weight and frontal structure, similar to light truck or SUV.

FIGURE 8.10 Illustrations of the side-impact devices in the 5-star safety rating system and IIHS's rating system (Not to scale).

To increase the safety challenge of the test, IIHS analysts put two small dummies in the driver's seat and in the rear seat behind the driver. With the higher impact position of the device, this test also assesses the potential head injury risk of small-frame drivers and occupants. In addition, the IIHS analysts take into account what happened to the vehicle and occupant compartment when assigning a safety rating.

The IIHS also has a roof strength test. If you recall, the industry argued against stronger roofs saying that they *would not* save lives. But the IIHS insisted that stronger roofs *would* save lives. When NHTSA upgraded the Roof Crush Resistance Standard, the IIHS asserted that the standard did not go far enough. The roofs, and therefore the standard's roof-crushing test criteria, should be tougher.

While the IIHS could not create a federal standard, they could create a tougher test within their safety rating system. And they did. To receive the highest safety rating, *Top Safety Pick*, vehicles have to have a roof-to-strength ratio of at least 4.0, i.e., stronger than the federal standard of 3.0. And manufacturers want that highest rating because they are using safety as a selling point. The IIHS created a tougher roof test as a carrot within their rating system to which the industry responded.

In addition to crash testing vehicles and crushing roofs, the IIHS safety rating system also assesses other built-in equipment such as head restraints, seats, and headlights, as well as driver-assist technologies to prevent frontal crashes with other vehicles and pedestrians. The IIHS tests and criteria for achieving the highest safety rating Top Safety Pick+ can change quite frequently. Nonetheless, the industry appears to rise to the challenge and covet these awards.

After nearly three decades, manufacturers now are showcasing their high safety ratings as part of their marketing campaigns. To use the labels *IIHS Top Safety Pick* or *Top Safety Pick+* in advertising, carmakers must make a request to the IIHS. Between 2006 and 2014, the IIHS approval of these requests increased from about 220 to over 600, signaling high demand from manufacturers. It appears that the consumer movement to *test safety into cars* that NHTSA started in the 1970s – and IIHS expanded in the 1990s – has now become a competitive advantage in the marketplace.

The two safety-rating systems described here use new vehicles and crash them in controlled, laboratory settings. Every element of the crash test procedure is pre-specified and followed, to the letter. But the IIHS also has another safety rating system – one that uses real, human fatality data.

DRIVER DEATH RATE SYSTEM

About every three years, IIHS analysts use fatality data from the Fatality Analysis Reporting System (FARS) and vehicle registrations to generate a different kind of safety-ranking system. The key metric is a vehicle's driver death rates. The method behind the rankings removes the influence of a driver's age and gender. The result is a driver death rate ranking that is gender and age-neutral. The IIHS compiles hundreds of vehicles into a list from the worst (highest death rate) to the best (lowest death rate). In addition, each vehicle receives four individual ratings (and consequently four rankings): an overall death rate, a death rate in single-vehicle crashes, a death rate in single-vehicle rollover crashes, and a death rate in multiple-vehicle crashes.

For example, the IIHS ranked more than 250 vehicles for model years 2015-2017. The worst ranked vehicles were the Ford Fiesta and Hyundai Accent, both 4-door, mini cars. Among the highest ranked vehicles were the large SUV GMC Yukon and mid-size, luxury SUV Infinity QX60.

In comparison to the federal 5-star safety rating system and IIHS *Top Safety Picks*, this rating system is very data intensive. It uses the real crash experience of the vehicles, rather than laboratory testing. However, a consumer needs considerable statistical knowledge to properly interpret the ratings. As a result, the system is not designed for use by the typical car buyer.

NOTE FROM THE AUTHOR

While the NHTSA 5-star safety rating system and the IIHS *Top Safety Pick* system have incentivized manufacturers to build safer vehicles, they do not go far enough in my opinion. I believe that it's time to update our concept of safety ratings.

First, these laboratory-based tests need to be more realistic. In particular, the frontal crashes should better reflect the real-life situations that car occupants face.

While I am not prone to using sports analogies, I think that it's useful here. Wrestling has distinct weight classes. There's an obvious reason you don't pit a 285-pound wrestler against a fellow weighing 125 lb.

We do not have the same luxury of choosing the weight and size of the vehicles on the road with us. Nor can we prevent aging or whether we are born female or male.

As such, the frontal crash tests need to become more inclusive. Two-vehicle frontal crashes involving small and large vehicles should be added to the tests. Dummies should be redesigned to better represent women and seniors. In these ways, the tests may better reflect the real hazards faced by car occupants.

Secondly, while the federal NCAP program could add these improvements to the 5-star safety rating system, realisitically, it would take years. In contrast, the program by the Insurance Institute for Highway Safety that generates the *Top Safety Picks+* has shown that it can adapt quickly with tougher tests using smaller dummies, a bigger device in the side-impact test, and a higher roof strength requirement.

As the insurance industry faces the aging population, I can envision IIHS adopting more senior-friendly dummies in their laboratory test crashes. And carmakers are likely to try to pass their new tests. After all, they love that top safety rating - we hear it all the time in their marketing ads.

In my opinion, NCAP should be drastically reduced or totally eliminated. Let's face it, when it comes to incentivizing the industry to make safer cars, IIHS does it better and faster.

Lastly, I believe that a vehicle's real-life safety performance is more valuable to the consumer than results from structured, laboratory tests. It's time to modernize our rating system to include more on-the-road data. Vehicles are meant to be driven by humans on the road. Why not judge their safety in the same way?

Herein lies the reason why I developed the safety rating system called Auto Grades available on www.TheAutoProfessor.com. Using crash data from the Fatality Analysis Reporting System, all vehicles are rated based on how well they protect drivers – women, men, young, and old – during chaotic events on the road. Safety should be defined by the data that tells us what happens to real people in real crashes, not dummies in controlled, laboratory tests.

NOTE

1. Claybrook, Joan & Advocates for Highway and Auto Safety. (2019). *NCAP at 40: Time to return to excellence.* Advocates for Highway and Auto Safety. https://saferoads.org/wp-content/uploads/2019/10/NCAP-at-40-Time-to-Return-to-Excellence-by-Joan-Claybrook.pdf.

CHAPTER 9

Two Recall Debates

Identifying unsafe vehicles and equipment and getting them
repaired or off the road is integral to saving lives, improving safety,
and reducing the costs involved with motor vehicle crashes.

*—NATIONAL HIGHWAY TRAFFIC SAFETY
ADMINISTRATION (2020)*[1]

Most family households today have at least two vehicles. Part of being
a vehicle owner is receiving a dreaded recall notice. In 2020, carmakers
recalled nearly 32 million vehicles, about 1 recall per four households.
Figure 9.1 shows how the number of recalled vehicles has increased dramatically since 2013.

But this common experience today outraged one car owner from Falls
Church, Virginia, in 1966. He sent the recall letter to Ralph Nader who, in
turn, gave it to then-Senator Walter Mondale from Minnesota.

Dear Buick Owner:

It has been called to our attention by Buick Motor Division that a bolt installation on the brake mechanism of your particular (1965) Buick LeSabre ...
which we delivered to you might prove to be **troublesome** [emphasis added]
sometime in the future.

In order to forestall this possibility, it would be appreciated if you would
bring your car into our Service Department in the immediate future in order
that we may check this installation and make any necessary corrections ...

DOI: 10.1201/9781003035343-9

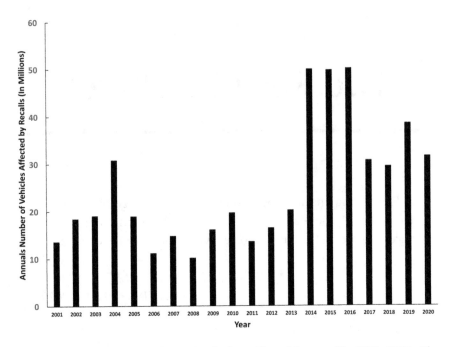

FIGURE 9.1 Annual number of vehicles affected by recalls 2001–2020. The General Motors ignition-switch recall and Takata air bag recall caused the spike in vehicles affected starting in 2014. In 2015, the Volkswagen vehicles with faulty emission systems added to the recall volume.

Mondale entered this letter into the U.S. Congressional Record on June 24, 1966, during the debate on the National Traffic and Motor Vehicle Safety Act. He explained that if the bolts in question did become loose, then a wheel on the car could fall off without warning. Mondale mocked the idea that this was merely "troublesome." He voiced concern about the fate of families that owned a 1965 Buick LeSabre.

The letter never said that the problem was a safety hazard. Rather than "conceal the nature of the problem and the danger involved," the young senator argued that the car industry had a duty. That duty was to clearly describe defects to car owners, to assess the safety risks, and to explain the steps needed to fix each flaw in a timely manner.

Mondale convinced his colleagues. Included in the final version of the act, signed by President Johnson in 1966, was a requirement that carmakers issue recalls on vehicles. But who would pay to fix these problems? It took many consumer advocates and years of legislative lobbying to require the industry to pay for the repairs. President Ford signed a 1974 bill that clearly shifted the financial burden to carmakers.

You might think about recalls as fixes "after the horse has left the barn." In other words, people use their vehicles in the normal course of their daily lives. Then, something happens. Maybe a tragedy occurs. Or, someone notices something about the vehicle and thinks, "this should not be this way." That someone may be a person like you. Or, that critique might come from a consumer advocacy organization.

JEEP FUEL SYSTEM DEBATE

Clarence Ditlow, executive director of the Center for Auto Safety (CAS), petitioned the National Highway Traffic Safety Administration (NHTSA) in October 2009 to open a defect investigation into the rear-mounted fuel systems of the Jeep Grand Cherokee, model years 1993–2004. His consumer watchdog group used the data from 172 crashes and 254 fatalities to claim that these SUVs had defective fuel systems. Furthermore, the CAS cited 64 fatalities where authorities identified fire as the *most harmful event*. In other words, police believed the fire caused the fatality, not the crash. As such, the fuel systems presented a fire hazard in collisions and should be recalled.

The Office of Defects Investigation (ODI) is the agency within NHTSA that studies possible defects and oversees recalls. On September 3, 2010, the ODI began investigating the Jeep fuel system.

> Sally was unaware of the defect inquiry on that fateful day in March 2012. She carefully strapped her little cousin Tommy into his car seat in the second row of the family's 1999 Jeep Grand Cherokee. Then, she drove her usual route through their small town in Georgia to Tommy's tennis lesson. She pulled into the left-turn lane and waited for the traffic to clear. Within minutes, a 1996 Dodge Dakota truck rear-ended the Jeep and it burst into flames. Tragically, Tommy died in the fire. The crash became part of the defect investigation.

We can never say for certain if this sad event influenced the decision, but within three months, the ODI upgraded its investigation and inched closer to a recall of Jeep's SUVs. The process would take another two years to complete. But even as the ODI considered punitive action that might undermine Chrysler and its future, other parts of the federal government were doing everything they could to save the company.

Like other U.S. carmakers, Chrysler was hit hard by the Great Recession of 2007–2009. The company filed for protection under Chapter 11 of the U.S. Bankruptcy Code on April 30, 2009. This was about five months before the recall process began. The Obama administration played a leading role in saving Chrysler, the third-largest U.S. automaker. By lending Chrysler more than $8 billion to restructure, the federal government ensured the company's continued operations. As documented in Chrysler's filings with the ODI during the probe, the newly formed Chrysler Group LLC assumed responsibility for safety recalls under the terms of the sale of the company's assets on June 10, 2009.

THE RECALL PROCESS

A tragedy, such as Tommy's, can motivate a car safety debate. The ODI's challenge is to determine if the tragedy was caused by a vehicle defect, according to their own criteria:

> By definition, a defect relates to motor vehicle safety when it involves an unreasonable risk of a crash occurring or an unreasonable risk of death or injury in the event of a crash.

The ODI uses risk-based processes in the four stages of their analysis, as shown in Figure 9.2. ODI analysts reach their conclusions by collecting data, reviewing it, doing an in-depth examination and, if warranted,

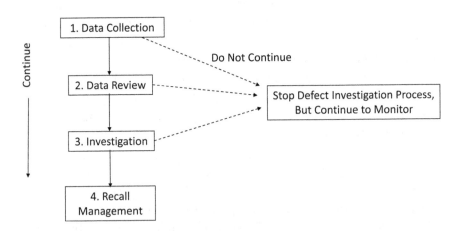

FIGURE 9.2   The steps in the defect investigation process. At any point in the process, decision-makers can choose to stop actively researching the problem.

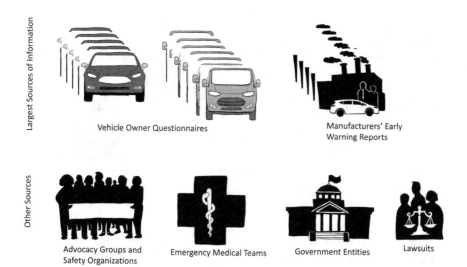

Largest Sources of Information

Vehicle Owner Questionnaires

Manufacturers' Early
Warning Reports

Other Sources

Advocacy Groups and
Safety Organizations

Emergency Medical Teams

Government Entities

Lawsuits

FIGURE 9.3 Examples of information sources in the data collection step of the defect investigation process.

managing the recall process. At each stage, they must decide whether to continue toward a possible recall or to essentially shelve the complaints as either not a safety issue or not backed by enough evidence to continue the process. Carmakers, consumers, and advocates watch and question each ODI decision.

The ODI receives information from a variety of sources during the first stage, Data Collection, as Figure 9.3 illustrates. Much of Data Collection is gathering complaints. Some complaints come directly from the vehicle owners, but they also can come indirectly through the manufacturer, advocacy groups (e.g., insurance agencies), emergency medical teams who have observed a problem (e.g., air bag did not deploy), government entities (e.g., lawmaker received a complaint from a constituent), or lawsuits (e.g., civic complaints about equipment failure).

If a consumer wants to complain directly to the ODI, then the vehicle owner questionnaire (VOQ) is the easiest way. In 2020, the ODI received more than 6,000 VOQs per month.

Alternatively, consumers can complain to their dealerships or carmakers. These complaints also are, by law, to be reported to the ODI through the early warning reporting (EWR) requirements. Volumes of other information, such as technical service bulletins (TSB) and other notices about defects, must also be submitted by manufacturers to the ODI. In 2019, the

Gather more detailed information from consumers, including event data recorders in the vehicles.

Use statistical methods to search for patterns in the data or predictions of future problems:
- text mining of the consumer complaints,
- data visualization,
- outlier analysis, and
- trend analysis.

Search federal crash databases for similar incidents:
- Special crash investigations,
- Fatality Analysis Reporting System,
- Crash Reporting Sampling System,
- Crashworthiness Data Systems,
- Crash Injury Research & Engineering Network, and
- National Automotive Sampling System.

Perform a risk assessment.

Ask manufacturers for information regarding specific crashes, other claims, or similar types of incidents.

Research insurance claims.

Query Carfax reports, vehicle auction sites, news media.

FIGURE 9.4 Examples of data analysis methods used in the data review step of the defect investigation process.

ODI processed 125,000 field reports and TSBs, together with more than 1,400 EWRs from carmakers.

At the second stage of ODI's process, Data Review, analysts use statistical methods to compile and evaluate the evidence of potential safety-related defects, as shown in Figure 9.4. These methods include text mining for keywords in the complaint narrative, graphical visualization algorithms for detecting patterns in the data, and trend analysis of the frequency and severity of the complaints, to name a few.

Data gathering and sorting activities culminate in a risk assessment. The purpose is to determine if there is a potential defect. Analysts use a ranking system for weighing the severity level of the alleged defect and the frequency of occurrence. Based on the results, the ODI decides whether to go to the next level. That is, to officially open the defect investigation.

JEEP FUEL SYSTEM RECALL DECISION

When the CAS sent its letter about the Jeep defect in October 2009, the petition spurred the ODI to collect more data. By August 2010, there was enough evidence to start investigating the fire hazard of the 1993–2004 Jeep Grand Cherokee fuel system. The ODI published their data findings:

- Nearly 3,000 occupants of these vehicles died in all crashes since 1992.

- Fifty-five died in 44 crashes where fire was the most harmful event.

- In rear impacts, there were ten crashes and 13 occupant fatalities with fire occurrence.

OPPOSING

The Jeep SUVs exceeded the Fuel System Integrity Standard.

The vast majority of incidents cited by NHTSA had speeds in excess of standard requirement.

Other vehicles had worse fire-related, rear-impact fatality rates than the Grand Cherokee and Liberty; they were not recalled.

The Jeep SUVs had more than: • 500 billion miles, • fires are extremely rare.	Statistical analysis showed no significant differences in fire rates.

CHRYSLER-JEEP

SUPPORTING

The Fuel System Integrity Standard is too weak.

The analysis should focus on the 44 occupants in rear-impact crashes with fire as the most harmful event.

Jeep SUVs have the highest rates of fire as the most harmful event. The only exception is the 1994-2004 Ford Mustang with a rear-mounted fuel tank.

The number of fires in comparison vehicles was too small to permit any valid statistical analysis.	The fire rate comparison should be to modern SUV vehicles, not old models and small cars.

CENTER FOR AUTO SAFETY

FIGURE 9.5   Data-related issues raised in the defect investigation of the rear-mounted fuel system in Jeep Grand Cherokee and Liberty SUVs.

- Of the 12 reports from consumers, ten were about fires or fuel leakage. Among these, there was one alleged fatality and nine alleged injuries.

- Post-crash fires were not overrepresented in data reported by Chrysler.

Having completed two stages of the process shown in Figure 9.2, the ODI was ready for the third stage, Defect Investigation.

The probe lasted nearly four years. During that time, the ODI, Chrysler, and the CAS would disagree about a wide cross-section of data-related issues, as summarized in Figure 9.5. Chrysler argued that the fuel system on the Jeep SUVs had passed the Fuel System Integrity standard. This was evidence that the fuel system was safe.

The CAS responded that passing the standard was not a reason to stop the inquiry. After all, the federal standard was too weak. There could still be a defect even if the Jeeps met the standard. In addition, history proved that recalls occurred independent of standards. For example, when Ford built the 1971–1976 Pintos, the fuel system standard did not include a rear-impact test. Yet when the ODI investigated reports of Pintos catching fire, its analysts hit the back of the Pinto in tests to determine the existence of a safety-related defect. The Pinto failed and it was recalled. The Jeep SUVs should be recalled, as well.

Chrysler pointed out that the majority of incidents cited by the ODI were high-speed crashes. They involved speeds in excess of the standard's requirement of 30 mph. Even though the rear-impact, fuel integrity standard had been updated in 2007 from 30 mph to 50 mph, Chrysler argued that pre-certified vehicles did not need to be held to this higher standard.

To counter this argument, the CAS focused on the cause of death in the SUV crashes. As background, the Fatality Analysis Reporting System (FARS) has a data field for the most harmful event in its database. Crash scene investigators or medical personnel code this information as the most likely cause of death. If they list fire as the most harmful event, then the fatality was due to the fire and not the impact of the crash. For example, 4-year-old Tommy only had a broken leg from the crash, but he died because of the fire. The CAS wanted the ODI to focus on crashes with fire as the cause of death.

Even in the alleged high-speed crashes cited by Chrysler, investigators coded the fire as the most harmful event. To support its argument, the CAS reanalyzed Chrysler's data. The analysis showed that the Jeep SUVs had the second highest rate of fire as the most harmful event. The car with the highest rate was the Ford Mustang with its rear-mounted fuel tank.

In their risk assessment, the ODI compared the fire rates of the Grand Cherokee and Liberty to similar-use, **peer vehicles**. The set of peer vehicles were GM S10 Blazer, Ford Explorer, Toyota 4Runner, Isuzu Rodeo, Isuzu Trooper, Mitsubishi Montero, Suzuki Sidekick, and Suzuki XL-7.

Chrysler argued that the ODI was using too few peer choices. Instead, a wider comparison including cars with rear-mounted fuel tanks, such as the Ford Mustang, should be included in the analysis. There were vehicles with higher fire rates than the Jeep SUVs. Yet, none of these vehicles had been or were being recalled. Hence, Chrysler vehicles should be treated the same.

The CAS contended that the wide set of comparison vehicles suggested by Chrysler was inappropriate. Vehicles made before 1993 were not the right peers. Nor were small cars.

Chrysler maintained that fires in their vehicles were quite rare. There were millions of these SUVs on the road. Together, they accounted for billions of miles travelled. As such, the fires the ODI was investigating were a tiny proportion of the whole. Statistical analysis would find no significant difference between the rate of fire in Jeep SUVs and other vehicles. Again, why should these SUVs be recalled?

The CAS dismissed Chrysler's statistical analysis claim. The number of fires in the peer vehicles was too small to do a statistically valid study.

The peer SUVs used for comparison were:
- GM S10 Blazer,
- Ford Explorer,
- Toyota 4Runner,
- Isuzu Rodeo,
- Isuzu Trooper,
- Mitsubishi Montero,
- Suzuki Sidekick, and
- Suzuki XL-7.

★ ★ ★ ★ ★

NHTSA

Grand Cherokee and Liberty rear-impact fatal and non-fatal fire rates were poor in comparison to peer SUVs.

The decision was based on 75 fatalities and 58 injuries with:
- 56 fatal fires,
- 28 non-fatal fires, and
- 6 fuel leak incidents.

Risk Assessment: Grand Cherokee and Liberty had the worst post-crash fires and fuel leaks rates. The only exception in the comparison set was the small volume Suzuki Sidekick. Post-crash fires and fuel leaks in these vehicles pose a "substantial risk to safety." Vehicles should be recalled.

FIGURE 9.6   Key data-related findings used by NHTSA in deciding to recall Jeep SUVs.

NHTSA, or more specifically the ODI, reached a decision on June 4, 2013 – nearly four years after the defect investigation process began. Chrysler should recall the Jeep Grand Cherokee model years 1993–2004 and Jeep Liberty model years 2002–2007. Figure 9.6 highlights the data NHTSA used to reach this decision. These Jeep SUVs had the worst fire and fuel leakage rates in comparison to the ODI's peer vehicles. The fuel tank design represented a "substantial risk to safety."

In response, Chrysler suggested a recall of about half of the vehicles identified by NHTSA. They would recall the 2002–2007 Liberty vehicles and the 1993–1998 Grand Cherokees, but not the 1999–2004 Grand Cherokees. Chrysler said that these later-year models had lower fire rates. In the end, NHTSA agreed to this smaller set of vehicles for recall. The vehicle Tommy died in, the 1999 Grand Cherokee, would not be recalled.

As part of the recall agreement, Chrysler would install a hitch receiver (tow hitch) on approximately 1.6 million affected vehicles. NHTSA accepted laboratory tests showing some incremental safety benefits of the hitch receiver in certain low- and moderate-speed crashes. For those SUVs not being recalled, Chrysler would monitor customer satisfaction.

Chrysler no longer builds vehicles with rear-mounted fuel tanks.

There is no doubt that fires occurred in these Jeep SUVs. Everyone could plainly see the charred wreckage. But what happens when car owners claim a safety-related defect that cannot be seen? A hazardous event occurs, but leaves no obvious trace.

A 2012 report by the National Research Council predicted this problem, due to the shift from mechanical systems in cars to electronic ones:

> Some failures of software and other faults in electronics systems do not leave physical evidence of their occurrence, which can complicate assessment of the causes of unusual behaviors in the modern, electronics-intensive automobile.

Tesla's all-electric vehicles have fulfilled this prophecy.

TESLA SUDDEN UNINTENDED ACCELERATION DEBATE

On May 6, 2019, a woman in Georgia complained to NHTSA. She described how she went from being a Tesla enthusiast to a disappointed owner. The cause of her disenchantment: an experience with sudden unintended acceleration (SUA). Figure 9.7 reproduces excerpts from her complaint. NHTSA assigned her complaint VOQ #11206155.

Brian Sparks, an investment advisor, repeatedly petitioned the ODI to investigate the problem of SUA in Tesla vehicles. Among the evidence that he submitted to the ODI were 232 complaints from the VOQ federal database. Complaint VOQ #11206155 in Figure 9.7 was part of his petition.

On January 13, 2020, the ODI started the SUA defect investigation process for the Tesla Model S, Model X, Model 3, and Model Y for the model years 2012–2020. That is, the probe covered all the years these vehicles had

1. On May 6th, sometime between 8:20 and 8:30 pm, I pulled into the driveway of my home and slowed to a halt to park my Tesla. I was returning from work after stopping to see a friend on the way.
2. As I was waiting for the garage door to fully open, the car suddenly and violently accelerated and lurched forward.
3. I immediately jammed the brakes but could not stop before the car hit the stone wall (separating our two car garages is a stone pillar) – mine was on the right side garage. My husband's Prius was parked in the left garage which was closed at that time.
4. I have attached pictures of the broken wall and my car. You will see the damages to the car on the right side of the car which are on the opposite side from the center stone wall. Even if I had pressed the gas pedal (as tesla [sic] is most likely to claim!) there is no way that I could have hit the right side of my car.

FIGURE 9.7 Excerpts from vehicle owner questionnaire (VOQ) #11206155. The owner described her experience of sudden unintended acceleration of her Tesla Model 3.

been made up until that point. NHTSA defined unintended acceleration as "the occurrence of any degree of acceleration that the vehicle driver did not purposely cause to occur."

Within a week of the ODI starting the process, Tesla published a blog on its website denouncing both Sparks and the allegations of SUA. The blog labeled Sparks a "short-seller." That is, if Tesla's stock went down – which could happen with news of a defect investigation – then Sparks would make money. The remainder of the Tesla blog contended that all complaints were "completely false."

As he later admitted, Sparks had taken a short position on Tesla stock at the time of his petitions. But he argued that this fact did not make the complaints less credible.

The ODI reviewed the data to determine if there was a possible safety-related defect in Tesla vehicles. Its analysts attempted to answer the following questions:

- What was the root cause of this potential, safety-related problem?

- Was it a vehicle-based cause that needed further examination and possible recall?

- Or, was it human error? Was the driver pressing the accelerator pedal when s/he thought it was the brake?

Table 9.1 shows a tabulation of all the cases considered. Nearly half were not investigated due to lack of data or untimely receipt of information. Of the 118 cases studied in detail by the ODI, the vast majority occurred in parking lots, driveways, or other close quarter, "not-in-traffic" locations. Analysts took statements from the drivers and collected information from the vehicles' electronic data recorders, log data from Tesla, and/or video data from Tesla. Each of these electronic data sources contained coded and/or encrypted information about the vehicle systems, i.e., what was happening at the time of the crash.

The inquiry lasted about one year. In that time, the ODI reviewed the incidents, as well as the service histories of the vehicles. Analysts also studied Tesla's systems in detail, including its safeguards for the accelerator pedal position sensor assembly, motor control system, and brake system design. Figure 9.8 summarizes the process the ODI followed and the data-related evidence it accumulated in reaching a decision to not recommend a recall.

TABLE 9.1   List of Incidents Considered by ODI Concerning Sudden Unintended Acceleration in Tesla Vehicles

Category	Data Reviewed	Data Not Available	Data Not Obtained	Total
Parking Lot	61	44	9	114
Driveway	26	16	4	46
Traffic Light	11	7	2	20
Parking Garage	7	5	1	13
City Traffic	3	1	0	4
Stop-and-go Traffic	2	2	0	4
Highway Traffic	2	1	1	4
Stop Sign	2	1	0	3
Charging Station	1	1	1	3
Street Side Parking	1	1	0	2
Drive Thru	1	0	1	2
School Drop-Off Lane	1	0	0	1
Car Wash	0	1	0	1
Gated Exit (China Incident)	0	1	0	1
	118	80	19	217

118 incidents of sudden unintended acceleration were reviewed.

Crash data reviewed includes:
1. event data recorder,
2. log data from Tesla, and
3. video data from Tesla.

Vehicles always responded as expected:
• Accelerated when accelerator pedal was applied,
• Slowed when the accelerator pedal was released, and
• Slowed more rapidly when brake was applied.

In 100% of incidents, driver error in pressing the wrong pedal caused the event.

There was no evidence of a vehicle-based cause.

In 90% of incidents, the driver did not brake.

In 10% of incidents, the driver did brake late in the event.

FIGURE 9.8   Key data-related findings used by NHTSA/ODI in deciding not to open a defect investigation of sudden unintended acceleration in Tesla vehicles.

In closing the defect investigation, ODI analysts cited incident VOQ #11206155 excerpted in Figure 9.7 and accepted Tesla's explanation that **driver error**, pressing on the accelerator, created the problem:

> According to the vehicle's diagnostic log, immediately prior to the incident, the accelerator pedal was released, regenerative braking was engaged and slowing the vehicle, and the steering wheel was turned to the right. Then, while the vehicle was traveling at approximately 5 miles per hour and the steering wheel was turned sharply to the right, the accelerator pedal was manually pressed and over about one second, increased from approximately 0% to as high as 88%. During this time, <u>the vehicle speed appropriately increased in response to the driver's manual accelerator pedal input</u> [emphasis added]. In the next two seconds, the accelerator pedal was released, the brake pedal was manually pressed, which also engaged the Anti-Lock Braking System, multiple crash-related alerts and signals were triggered, and the vehicle came to a stop.

The ODI further asserted that correct interpretation of the data from the electronic data recorder supported their assessment – that the SUA came from the driver pressing the accelerator pedal.

In the end, the ODI arrived at the same conclusion for all of the complaints it reviewed. All the owners were wrong. Without exception, the cause of the SUA in the Tesla's was human error. In addition, the drivers in 90 percent of the incidents never applied the brakes, according to the ODI.

On January 8, 2021, the agency stopped its defect investigation process. Yet, as this manuscript goes to print, Tesla owners continue to post complaints about SUA on the NHTSA website.

NOTE FROM THE AUTHOR

The data needed to diagnose car defects in the future will be markedly different than data used in the past. The ODI collected information for studying fires in Jeep SUVs from accident reports, first responders, and other crash scene investigators. In the future, identifying electronic-related defects will require a totally different type of data.

The Tesla corporation is leading the industry in electronics, data collection, and software. As such, the experience in investigating Tesla vehicles provides insight into the challenges facing the car safety community.

Future defect and recall investigations will depend more heavily upon vehicles' electronic data recorders, company-collected data, and proprietary software. Just as Tesla vehicles are in constant contact with the "mother ship," future electronics-laden vehicles will be connected to their manufacturers and even other vehicles, through wireless communications. These data streams will be critical to identifying safety hazards. All regulators, both here and abroad, are facing the same challenges, with limited success.

Investigators need access to these data. And they need the knowledge and experience to independently analyze company-collected data and proprietary software. This independence is key to the ODI's relevance in safeguarding the driving public.

NOTE

1. National Highway Traffic Safety Administration. (2020). *Risk-based processes for safety defect analysis and management of recalls* (Report No. DOT HS 812 984). https://www.nhtsa.gov/document/risk-based-processes-safety-defect-analysis-and-management-recalls

CHAPTER 10

The Automated Driver-Assistance Systems Debate

In the past, automotive safety technologies focused on protecting drivers and passengers after a crash. Technology is evolving to the point that crashes can be largely prevented. Automated vehicle technology can drastically reduce and potentially eliminate the 94 percent of crashes caused by human error.

—*INTELLIGENT TRANSPORTATION SOCIETY OF AMERICA (2021)*[1]

In the mid-1970s, a research team at Indiana University set out to find the cause of car accidents. Using a sample of urban crashes, they found that "93% of accidents were definitely or probably caused by the driver." A study in Europe obtained similar results. The 93% rate stood largely unchallenged for nearly 30 years.

With new driver-assistance technologies and the growing interest in self-driving (or autonomous) vehicles, the National Highway Traffic Safety Administration (NHTSA) updated the study for the United States in 2005. Using more resources than before, its researchers constructed an extensive, nationally representative sample of crashes. However, in the end, the results did not change much. The new study found that drivers played a role in events leading up to a crash 94% of the time.

DOI: 10.1201/9781003035343-10

The researchers cautioned against saying that the driver *caused* the crash, as the quote at the beginning of this chapter does. That caveat is because the data may be incomplete. For example, suppose a deer darted across a road and caused a driver to swerve into a tree. The deer is long gone, and the driver is dead. In that case, the deer caused the crash. However, without witnesses, the investigators *do not know what caused* the crash.

Knowing that there may be some unrecorded factors that caused a crash, the NHTSA analysts in 2005 were more comfortable saying that "drivers' actions or inactions led" to the crash 94% of the time. In our example, while the driver's action of avoiding the deer did result in the crash, the cause of the crash was the deer. More generally, the types of driver errors found in the research were:

- not seeing another car or pedestrian

- driving too fast

- distraction

- drinking or drugs

- misjudging the speed of other vehicles

- falling asleep at the wheel

Drivers make mistakes. That's nothing new. Wasn't this known back in the early 1900s when the drivers were testing the limits of the Model T? Isn't that why we get traffic tickets? We drive too fast and make other bad decisions. To err is human and even the brightest minds make mistakes.

In 1997, a series of six chess matches between World Chess Champion Garry Kasparov and an IBM computer named Big Blue drew the world's attention. Up until then, humans could outsmart even the most sophisticated computer programs. The world watched while Kasparov confidently won the first game, but then he lost the second. The next three games each ended in a draw. In the last game, Big Blue shocked both its opponent and the world when it decidedly won the match.

Controversy followed. Some observers, especially Kasparov, believed that Big Blue's moves on the chess board were too human to be solely the machine. They thought that a real grandmaster must have been behind the machine's moves. No artificial intelligence could be that human in complex thinking, or that clever in resisting the temptations that Kasparov used as part of his game strategy.

The truth finally came out when IBM dismantled Big Blue and released its log and code. Both the machine and Kasparov had made some mistakes. But Kasparov did not take advantage of the machine's errors. At the time, the most important finding was that there was no hidden expert dictating the moves. The machine had beat one of the most developed human brains in a competitive, intellectual matchup.

Yes, both man and machine made mistakes. However, the machine turned out to be the better player. Looking back, the computer win changed the world's view of the power of artificial intelligence. Simply stated, *machines will be smarter than humans at some point in the future.* And if this is true, we should use computers to solve the most complicated problems facing humans. That was the thinking in 1997.

In the quarter of a century since that matchup of man versus machine, it is commonplace for computers to do both manual and intellectual tasks that improve or influence our lives. Computer algorithms diagnose medical ailments, trade in the global financial markets, predict tornadoes, and direct air traffic, just to name a few. And on a more personal note, they answer our questions ("Alexa, what time is it in Singapore?"), recommend music, and tell us to exercise more. We welcome computerized assistants, from front door cameras that tell us our vitamins just arrived to robotic discs that vacuum up cat hair in our homes. Their purpose is to improve our lives.

As such, it seems only logical to ask how computers could help keep us safe on the road. If human errors contribute to 94% of crashes, then the natural question is how could computers help to prevent these errors?

OVERVIEW

Enter automated driver-assistance systems (ADAS). Table 10.1 lists some modern ADAS. These are small computer systems designed to solve a certain problem. Brake for a pedestrian. Slow down on a curve. Dim lights with oncoming traffic. These systems *see* a problem, *suggest* a solution, and if necessary, *take action.* For example, if a car in front of you is stopped, and you do not slow down, then the forward-collision warning system signals you to start braking. If you still do not act fast enough, then the

TABLE 10.1   Automated Driver-Assistance Systems Available in 2021

• Adaptive Cruise Control	• High-speed Alert
• Adaptive Headlights	• Hill Descent Assist
• Anti-lock Braking System	• Hill Start Assist
• Automatic Emergency Braking	• Lane Departure Warning
• Automatic Parallel Parking	• Lane Keeping Assist
• Back-up Camera	• Left Turn Crash Avoidance
• Back-up Warning	• Obstacle Detection
• Bicycle Detection	• Parking Sensors
• Blind Spot Detection	• Pedestrian Detection
• Brake Assist	• Rear-cross Traffic Alert
• Curve Speed Warning	• Temperature Warning
• Drowsiness Alert	• Tire Pressure Monitoring System
• Electronic Stability Control	• Traction Control
• Forward-Collision Warning	

automatic emergency braking system rapidly stops the car. You avoid a crash and you don't have to call your insurance agent. It's a happy day.

Now if ADAS could solve specific problems behind the wheel, why not link all these smaller systems and solve the bigger problem – the human driver? Why assist humans when computers are smart enough to do it all? That describes the mindset of those who advocate for autonomously driven vehicles. Eliminate the driver, eliminate errors, eliminate crashes. What will it take to realize this vision? What will it take to build these robots, train them to drive on our roads, and dispatch them throughout our city streets, country roads, and highways? As Chapter 11 discusses, lots and lots of data.

But in this chapter, the focus is on the role of data in the evolution and debate surrounding one automated driver-assistance technology: the forward-collision warning system. Its story highlights the key role of data in identifying safety problems and the potential benefits of ADAS. It also shows the drawn-out process of making such technologies standard on our cars. For at least one family, this advance might have saved them a lifetime of loss.

No parent wants to bury their child. Especially not at the tender age of 4 years old.

Vanessa was driving west on State Route 101 in Phoenix in evening traffic. It was August 2015, and she could see some thunderstorm clouds gathering up north. Vanessa stopped her sedan behind another vehicle near an exit ramp.

In the rear-view mirror, she had a clear view of her daughter Emma, who was buckled in a front-facing car seat sitting directly behind the driver's seat. That's the last time Vanessa saw her daughter alive.

An SUV struck them from behind. The force of the crash collapsed the trunk of the sedan and wreckage pushed into the back seat. It also caused Vanessa's seatback to collapse backward into Emma.

The SUV driver was Jackie, a nurse who was almost home after a long day at work. Still in her scrubs, Jackie told police the crash happened so fast. She saw brake lights and then felt the impact. It was Jackie who administered CPR to Emma until paramedics arrived on the scene.

A forward-collision warning system in the SUV may have provided Jackie with the extra seconds she needed to brake – and spared Emma's family the agony of burying their daughter.

AN EARLY DRIVER-ASSISTANCE SYSTEM

Rear-impact crashes make up about one third of all crashes. George Rashid, a Lebanese immigrant trained in electronics and owner of three car dealerships, saw the value of preventing these types of crashes. Using his training, Rashid worked for three years to create a way to prevent rear-impact crashes. In 1957, he obtained a U.S. patent for an *automatic vehicle control system* that stops or slows the vehicle in dangerous conditions. The purpose of this forward-collision avoidance system was to alert the driver to the imminent risk of a crash using a series of beeps. If the driver ignored the warning, the device would use vacuum tubes to brake the vehicle.

More than three decades later, NHTSA put together a team to find out how well Rashid's invention actually worked. The team concluded that the system provided useful warnings of imminent collision, so long as the closing speed with the vehicle ahead was not too great – typically, between 10 and 30 mph. The system was most effective in warning the driver during merges onto busy highways, when the driver was looking over her/his left shoulder and the vehicles in front of them were slowing. In summary, this early evaluation was positive.

In 1991, the U.S. Department of Transportation formed the Intelligent Transportation Systems Joint Program (ITS JP) office. Its mission? "To improve the safety and mobility of people and goods through collaborative and innovative research, development, and implementation of intelligent transportation systems." One of its first major tasks was to assess the benefits of some ADAS: forward-collision warning systems, lane change/merge guidance systems, and road departure alarm systems. By the mid-1990s, sensors, processors, control systems, and digital displays had advanced to a stage that made these driver-assistance technologies cost effective and highly reliable.

Using data from a nationwide sample of crashes, the investigators estimated that three types of crashes were responsible for more than 3 million crashes in 1994, as shown in Table 10.2. This represented *nearly half of all crashes on our nation's roadways.* Together, the economic cost of these crashes was nearly $70 billion annually, including the lifetime cost of fatalities, injuries, and vehicle damage.

However, the researchers also knew that these ADAS could not prevent all relevant crashes. For example, a vehicle might move too fast toward a stopped vehicle for a forward-collision warning system to alert the driver to slow down. Or, there might be a blizzard or fog that reduced the ability of the system's camera to spot danger. Nonetheless, this early research showed that there were clear benefits to the ADAS.

The team estimated that ADAS could reduce these three types of crashes by 51% (rear impact), 47% (lane change/merge), and 65% (road departure). Those reductions would result in over a million crashes avoided annually. Clearly, this was a goal worth pursuing.

Then the team looked at the projected financial impact. If all vehicles had these three ADAS, and they were effective at the rates listed in Table 10.2, the NHTSA researchers concluded that the nation would save $25.6 billion a year. That's over a third of the $69.6 billion shown in Table 10.2.

The final takeaway from the study was a cost benefit analysis. That is, if manufacturers could keep the cost of the three ADAS below the break-even number of $1,500 per vehicle, then society would save lives and money.

By 1996, two NHTSA studies had established the benefits of ADAS, in general, and forward-collision warning systems, in particular. The data added up to a positive outlook for these new technologies. However, the industry appeared unresponsive. Perhaps, carmakers worried about whether customers would accept these advances.

TABLE 10.2   Number, Economic Costs, and Potential Savings of Three Types of Crashes in 1994

Type of Crash	Estimates				
	Number of Crashes in 1994, in Millions	Economic Costs in 1994, in Billions	Potential Percentage Reduced by Driver-Assist Technology	Potential Number of Crashes Reduced	Potential Savings If All Vehicles Had These Technologies, in Billions
Rear Impact	1.66	$35.4	51%	791,000	
Lane Change/Merge	0.24	$3.5	47%	90,000	
Road Departure	1.24	$30.7	65%	297,000	
Total	**3.14**	**$69.6**		**1,178,000**	**$25.6**

EARLY IMPRESSIONS

Shortly after publishing their findings of the three proposed ADAS, the ITS JP office asked an important question about car sales. If the technologies were available, would people buy them? The office hired Charles River Associates, a consulting firm, to find the answer.

The researchers conducted focus groups with potential car buyers to gauge their level of enthusiasm for ADAS. Table 10.3 lists the safety problems and the ADAS studied.

Moderators asked participants to make certain assumptions about these future technologies. First, they should assume that the ADAS would be "available at a price level that's quite acceptable to you." Second, that

TABLE 10.3   Automated Driver-Assist Systems Included in Early Study of Consumer Acceptance

Safety Problem	Automated Driver-Assistance Systems
Rear-object Crashes	Back-up Warning Devices
Run-off-the-road Crashes	Lane Trackers
Lane Change/Merge Crashes	Side-object Detection System
Rear-end Crashes	Front-object Detection
Drowsy Drivers	Driver Monitoring Systems
Vision Under Degraded Conditions (e.g., darkness, poor weather, or glare)	Vision Enhancement Systems
Intersection Crashes	[No system proposed during the 1997 study]
Highway Driving	Adaptive or Intelligent Cruise Control

warning devices such as beeps or bells would be "chosen to be effective for the majority of drivers."

The responses from these focus groups revealed how a consumer in the late-1990s might react when seeing a new car with these technologies in the showroom. Among the findings were:

- The rear-object or back-up detection and camera system was the most popular among the participants, especially among older drivers.

- The second most favorable ADAS was the side- or blind-spot detection.

- To a lesser degree, but still positively viewed, were the forward-collision and run-off-the-road warning systems.

- Some participants expressed concern that a car with all these ADAS might have "too many lights and sounds."

- Overall, most respondents were "favorable or enthusiastically disposed to the idea, though somewhat skeptical about whether in practice (the technologies) would work as advertised."

The other technologies listed in Table 10.3 received limited positive reviews. Overall, it was clear that educational campaigns would be needed before car buyers would embrace the new devices.

By the end of the 20th century, ADAS technology was maturing. Researchers who studied the worst car crashes strongly recommended immediate adoption of forward-collision warning systems. The technology could prevent deaths and injuries. In their opinion, it was time for the industry to step up and commit to making these systems standard equipment – even if car buyers were still on the fence.

PRESSURE APPLIED

The National Transportation Safety Board (NTSB) is often called to study the crashes of various modes of transportation, e.g., cars, trains, or airplanes. Its mission is to determine the cause of a crash and what might have prevented it. From the start of 1999 to the end of 2000, the NTSB investigated nine, rear-impact vehicle collisions. These crashes killed 20 people and injured 181 others. Though the details of each crash were different, the drivers had one thing in common. They were unable to detect slowed or stopped traffic in time to brake their vehicles and prevent a rear-end collision. In their 2001 report, the NTSB investigators concluded that one or more of the ADAS "would have helped alert the drivers to the vehicles ahead, so that they could slow their vehicles, and would have prevented or mitigated the circumstances of the collisions." Similarly, a 1992 study by Daimler-Benz found that if drivers had *an extra second* to react, 90% of rear-end collisions would be prevented.

At the end of their 2001 report, the NTSB team strongly recommended that NHTSA make forward-collision warning systems on passenger cars a federal mandate. They wrote that this standard should be backed by a public education campaign to win car buyers' acceptance.

When the NTSB issued their recommendation, only a handful of high-end vehicles had this technology as standard equipment. Carmakers follow a typical process to introduce technologies into their fleet of vehicles. First, they offer a new device as optional on the most expensive, new cars. After a few years, the equipment becomes standard on these luxury vehicles. Next, the process repeats on less expensive vehicles. First optional, then standard. This process usually takes about ten years to cycle through a manufacturer's fleet. And it takes about 15 years for the technology to become standard equipment across all new cars.

However, the industry's adoption of forward-collision warning systems did not follow this pattern. After 15 years, *less than* 5% of all cars had these life-saving systems as standard equipment.

NHTSA set out to change that in the fall of 2015. The automotive industry faced a choice: Either there would be a new federal regulation or manufacturers could voluntarily make rear-impact prevention technology standard on all future car models.

NHTSA had the evidence to support their edict. According to the latest crash data in 2011, about 53% of the 1.7 million rear-impact crashes would have been prevented or the harm reduced with this ADAS. The 53% was in line with the 1994 estimate of 51% shown in Table 10.1. Furthermore, NHTSA researchers projected that emergency braking with forward-collision warning systems would save 100 lives and prevent about 4,000 serious injuries each year. The Insurance Institute for Highway Safety (IIHS) also provided evidence of cost savings.

After six months of discussions, 20 major automakers signed a voluntary agreement in March 2016. Their automatic emergency braking systems would include a forward-collision warning system. And if the driver did not react fast enough, the vehicle would automatically brake. The timetable for smaller vehicles, those less than 8,500 lb gross vehicle weight, was set for model year 2022. The heavier vehicles, those between 8,501 and 10,000 lb gross vehicle weight, had until model year 2025 to conform. By model year 2021, nearly 95% of new vehicles in the smaller weight class had this braking system. There also was progress toward meeting the target for the larger vehicles ahead of schedule.

ASSESSING THE IMPACT

Much of this chapter described the slow journey of implementing just one of the available ADAS, forward-collision warning. For each of the systems listed in Table 10.1, a similar story could be told.

Experts forecast that vehicles equipped with these devices will help our society avoid crashes, save lives, reduce injuries, and save money. And the latest available data reveal that we could use the help:

- Nearly 100 people died daily as a result of motor vehicle crashes in 2020.

- 2.5 million drivers and passengers were injured and required emergency room treatment in 2015.

- Occupant injuries and deaths cost the United States over $75 billion in medical bills and productivity losses in 2017.

It will be many years before there are enough data to prove that the forecasted benefits of ADAS were correct. However, the IIHS released some early numbers that show we already are reaping rewards. Figure 10.1 illustrates the findings.

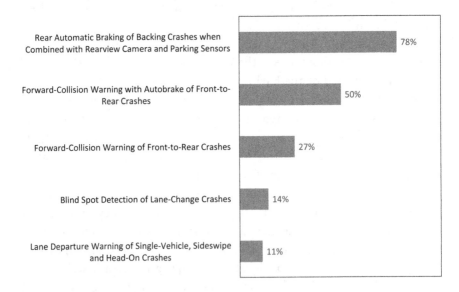

FIGURE 10.1 Estimated percent decrease of crashes with automated driver-assistance systems.

Using historical crash and recent insurance claim data, the IIHS found that forward-collision warning with autobraking had *halved* front-to-rear impacts. Other systems also have demonstrated substantial reductions in crashes, as shown in Figure 10.1. Bottom line: The automated driver-assistance systems are delivering on their promise.

NOTE FROM THE AUTHOR

In my opinion, the story of the slow pace of making ADAS standard on our vehicles is one of the greatest failures of the industry and our government regulators. All the evidence indicates that ADAS work. They reduce the reaction time of older drivers and nudge the distracted driver to react. They are proven to protect some of the most vulnerable road users, such as pedestrians, cyclists, and rear-seated children. Together, they correct for human failings and prevent crashes.

In Chapter 8, I mentioned that the European safety rating system, known as Euro NCAP, incorporates many of these technologies into their system. In essence, that system is using ratings as a carrot to sway carmakers to make ADAS standard on their vehicles. The IIHS is using the plus symbol "+" with their *Top Safety Picks* to signal these advancements. However, only now is the U.S. New Car Assessment Program considering rating some ADAS together with the 5-star safety rating sysem. In my view, this is too little too late.

At the outset of this chapter, I suggested that existing ADAS were building blocks for the self-driving vehicle. Nearly every week, another innovation adds to the excitement for a future without humans behind the wheel. Among the latest hot topics is vehicle-to-vehicle (V2V) communications. For example, when two vehicles are V2V equipped, then a driver is warned that it is unsafe to enter an intersection due to another cross-traffic vehicle. That is, unlike other ADAS, V2V technology does not rely on line-of-sight to anticipate a safety hazard. It can *see* around the corner!

This heightened enthusiasm has come with billions of dollars invested in the autonomous vehicle. However, progress in removing the human driver has been slower than some hoped. It turns out that the task of driving is more complex than connecting a series of small computer systems or winning a chess tournament.

There are many ways to discuss the advancement of self-driving vehicles. In Chapter 11, I focus on data's vital role in this emerging industry. Data are at the heart of the safety debate. Data create the robotic drivers that are designed to replace us. And data signal when these robots underachieve. After years of leading from behind, data's role is moving from the backseat into the driver's seat.

NOTE

1. Intelligent Transportation Society of America. (2021). *Safer: Advancing the research and deployment of ITS to save lives.* Retrieved October 4, 2021, from https://itsa.org/s/safer

CHAPTER 11

The Self-Driving Car Debate

Today we authorize @cruise to provide "driverless" autonomous
vehicle passenger service in which members of the public can ride
in a test vehicle that operates without a driver in the vehicle.

—*CALIFORNIA PUBLIC UTILITIES COMMISSION (2021)*[1]

It was supposed to be a carefree Saturday. Pamela Hesselbacher was
planning to eat dinner with her husband after walking their two
kids home from a park in Chandler, Arizona in November 2016.

Pamela pushed her infant daughter in a stroller and kept her
3-year-old son close as he rode his green bike. At a busy intersection, the family waited for the "walk" signal. Once things seemed
safe, Pamela stepped into the crosswalk.

Suddenly, a gray pickup came out of nowhere. The impact killed
Pamela and seriously injured her two children. The male driver
was sober, but still ignored the red light.

In March 2021, researchers at Waymo, an autonomous vehicle company
and subsidiary of Alphabet, published the results of a study looking at how
its technology might perform in simulated versions of real-life crashes
that occurred on the roads that it drives on in Chandler, Arizona, between
2008 and 2017.

DOI: 10.1201/9781003035343-11

The study reconstructed crashes in simulation, and then replaced simulated vehicle actors with its autonomously driven vehicle technology. First, the researchers examined how the technology would perform in place of the vehicle that initiated the crash. Then, the investigators looked at the same crash with autonomous technology in the responder role. The crash that killed Pamela Hesselbacher was among the 72 studied.

According to the simulations, when the company looked at how its technology would perform as the initiator, it noted that it would have avoided 100% of the crashes – including the crash that killed Pamela Hesselbacher. These findings suggest an autonomously driven vehicle would have stopped at the red light and "watched" the Hesselbacher family safely cross the intersection.

The primary motivation for building self-driving vehicles is to eliminate human error. Waymo's simulation study was more evidence from the industry that machines would perform better than human drivers.

Chandler was a logical choice for the study because the city is part of the 50-square-mile area used to test and train Waymo's vehicles. Figure 11.1 is a picture I took in November 2021 of one of these vehicles, commonly seen on the roadways. In fact, Waymo has been testing their vehicles since

FIGURE 11.1 A photo of an autonomously driven Waymo vehicle stopped at a traffic light in Tempe, Arizona taken on November 7, 2021 by the author. The fact that there were no humans in this vehicle, either behind the wheel or as passengers, appeared to attract little attention from other drivers.

2017 in the Metro Phoenix area – now deploying them without a human driver behind the wheel. These on-the-road miles are part of the process for qualifying a vehicle to operate robotically.

Waymo is one of dozens of companies that are attempting to make this vision of driverless cars a reality. Each company has its niche market. For example, currently there is pizza delivery in Houston by Nuro, commuter rides in San Francisco by Cruise, and long hauling on freight trucks by TuSimple in Tucson, Arizona. They also have their preferred terminology. Some use driverless, self-driving, or robotic vehicles, while others prefer autonomous driver or autonomous technology. Nevertheless, they are all striving for the same goal: a vehicle that operates without a human driver.

Every company also is following a similar, step-by-step process for building their self-driving algorithms for their computer systems. These algorithms or sets of rules, as you will see, are based on data. Data teach the algorithms. Data test the algorithms. And in the end, data determine if the robots are ready to be deployed. In summary, without data there would be no future for self-driving vehicles.

LEVELS OF AUTOMATION

As Chapter 10 discussed, there is no doubt that driver-assist technologies help reduce crashes. This progress is noteworthy. But we are still a long way off from eliminating human drivers altogether on our roadways.

Engineers and computer scientists, working within the Society of Automotive Engineers, have a classification scheme called the **Levels of Automation**. They measure progress toward self-driving cars using levels 0 through 5. Each level specifies the amount of driving done by the human versus the automated system. With each higher level, the human does less and the computer does more.

Level 0
Computers Advise,
Humans Drive **Level 0** means that a human performs all driving functions. The vehicle may provide a warning signal, such as a series of beeps in forward-collision warning systems. Yet, the driver is responsible for reacting to those beeps, i.e., applying the brakes.

Level 1
Computers Do Some Driving,
Humans at the Wheel At **Level 1**, computers built into the vehicle perform some driving tasks. Emergency braking is one such activity. Another example is adaptive cruise control. This device exhibits some *intelligence* in managing the driving function by maintaining a constant speed and distance from the vehicle ahead. However, the human

driver's hands are on the wheel, controlling where the vehicle goes in both Levels 0 and 1.

To reach the next levels of automation, steering control needs to shift from the human to the machine. But that transition is proving to be a challenge, as this example illustrates:

> A Tesla Model X was travelling south on United States Highway 101 in Mountain View, California in late March 2018. Police said the sport utility vehicle then steered into a highway gore area – a triangular marking located between lanes of the highway that indicates an exit ramp – and hit a crash attenuator (barrier) at 71 mph. After this initial impact, the SUV also hit a 2010 Mazda 3 and a 2017 Audi A4. The Model X driver, a 38-year-old man, was the only person to die in the crash. Performance data downloaded from the vehicle revealed that he was using the Tesla Autopilot, including traffic-aware cruise control and autosteer, lane-keeping assistance.

The National Transportation Safety Board (NTSB) investigators concluded that:

- The Tesla Autopilot system steered the SUV into the highway gore area due to system limitations.

- The driver did not adequately respond to the system failure due to distraction, likely a cell phone game, and overreliance on the Autopilot.

The NTSB also found a contributing factor in the crash: the SUV's "ineffective monitoring of driver engagement, which facilitated the driver's complacency and inattentiveness."

Level 2
Computers Drive More, Humans at the Wheel & Provide Backup — The Tesla Autopilot is a **Level 2** system. That means the driver has the role of monitoring what the automated system is doing. The computer system needs certain guidance because it does not recognize all situations. The driver can release some of the driving functions to the automated system, but she or he still needs to pay attention and be ready to take over.

The NTSB's mission in this type of crash is to learn what went wrong and to make safety recommendations for the vehicle's self-driving features.

But the job of parsing through the crash data is not as straightforward as it is with human-driven, more mechanical vehicles. Each self-driving vehicle is built by a company using proprietary software and hardware. There are different cameras, different settings on these cameras, and different algorithms, just to name a few. That is, the data or lessons learned from a crash are company specific and may not be relevant to other vehicles. Tesla is a prime example.

For example, the NTSB devoted nearly half of its 24-page report on the Mountain View crash to describing the functionality and limitations of the driver-assist programs and Autopilot system on the Tesla Model X. The second half of the report describes excerpts of data recorded about the crash. Most of the crash data came from the vehicle's electronic data recorder and the proprietary vehicle logs. The findings of this individual crash were based on these company-specific data sources.

The NTSB probe concluded that the Autopilot mistakenly interpreted the gore line markings as lane markings and steered straight into the gore. Tesla engineers did not design the forward-collision warning system nor the automated emergency braking to function at the 71-mph speed of the crash. According to the investigation, the Model X driver told family and friends that his Autopilot had, on prior occasions, steered into this gore. He was annoyed that he had to assume control and steer the vehicle back into the appropriate lane. The vehicle logs recorded these prior incidents. But on this fateful day, the driver did not take control in time.

As a result of this tragedy, Tesla made improvements to its proprietary systems. Engineers and computer scientists at other autonomous vehicle companies also studied this crash. After all, in an industry as young as this, any problem encountered by one company could be a lesson for all.

REACHING THE NEXT LEVEL OF AUTOMATION

Because the Tesla Autopilot functions as a Level 2 system, a court in Germany ruled that Tesla cannot use the term *autopilot* in its marketing there. The court reasoned that the term would give drivers false confidence in the system. In other words, human drivers likely would be inattentive, rather than ready to take over the driving function when needed, as expected.

So, does this expectation run counter to our nature? A 2016 survey by State Farm Insurance suggests the answer is yes.

State Farm researchers asked 1,000 drivers how likely they would engage in certain behaviors if their semi-autonomous (or Level 2) vehicle was driving itself. The researchers asked the participants to check all the categories that would apply to them, but to keep in mind that they might need to take over the wheel. The respondents said that they would be more likely to eat (48%), read text messages (45%), send text messages (43%), take pictures (36%), access the internet (36%), attend to children in the back seat (32%), make videos (26%), watch a movie (21%), or read a book (19%). Not surprisingly, our natural tendency is to engage in other, more enjoyable or distracting activities, rather than pay attention to the road ahead.

Herein lies a major problem. The time needed to disengage a driver from eating, texting, or other activities, and to re-engage them in taking control of the wheel is critical. It could be the difference between a crash occurring or not. This life-threatening delay can happen in a Level 2 or Level 3 automated vehicle.

An Uber test vehicle, in the form of a modified 2017 Volvo XC90 SUV, was navigating along an established test route in Tempe, Arizona, in March 2018. A female operator sat behind the wheel, but she was not in control. A pedestrian pushing a bicycle outside of a crosswalk moved into the Volvo's path. The pedestrian died in the crash. At the time of the incident, the vehicle was driving itself. Cameras inside the vehicle revealed that the operator was looking at her cell phone, which was mounted below the SUV console.

Level 3
Computers Drive Much More,
Humans at the Wheel &
Provide Backup

The Uber test vehicle had **Level 3** automation. At that level, the vehicle has what you might consider a "driving permit." The automated system does the driving, typically on a specific route. However, there is a human seated in the traditional driver's seat. That person should be ready to take over when the system gets into trouble.

In the Uber case, the NTSB findings revealed that the self-driving algorithm did not initially recognize the pedestrian or anticipate her walking path. The emergency braking system had been disengaged by an engineer. Instead, when the system realized that the vehicle was in a hazardous situation, it alerted the female operator. But she either did not respond or have enough time to avert the crash. In other words, the system and the driver failed.

Waymo wrote about this handoff problem in its 2021 safety report. As a vehicle's computer system takes on more driving tasks, then the risk of driver distraction increases. For this reason, Waymo decided to skip Levels 2 and 3 and work to develop a completely, self-driving vehicle. What will that take?

MORE AND MORE DATA

Level 4
Computers Do All the Driving,
Humans Monitor Remotely At **Level 4,** all the occupants in the vehicle are passengers. There is no steering wheel. The computer systems perform all the driving tasks. However, technicians give the vehicle a route and monitor its driving from a remote site. For example, tests of these types of vehicles are taking place at the University of Michigan and near Las Vegas airport.

Level 5
Computers Do All the Driving For a vehicle to be fully autonomous, or **Level 5,** it must be able to perform everything that you or I would do as drivers, including determining the route to arrive at a destination. Envision sending a message to your Level 5 car and telling it to pick up your takeout dinner from the local diner. The vehicle would decide the route to the restaurant, pull up curbside, receive the hot meal, and return home, i.e., a robotic delivery service.

But there is a lot that needs to occur before this dream is a reality.

To become competent drivers, we went through a process that trained us in the rules of the road and the functions of our cars. We studied how to interpret road signs, e.g., STOP, Yield, and School Zone. Our parents yelled at us or our driver education teacher suddenly assumed control of the wheel, often frightening us. However, these scary situations taught us how to control the car in different situations. All this training prepared us to take an exam. If we passed, then we could drive on a probationary basis. During that period, we gained real experience. And with experience, we earned the unrestricted driver's license.

Ample driving experience also is needed to remove the human (or in industry terms, the "meat") from behind the wheel and put in its place a computer algorithm (sometimes called the "silicon"). That experience comes in two stages. First, the self-driving algorithm needs to be trained in a computer laboratory. Next, it needs to be tested on the road.

The laboratory training uses millions and millions of examples of ordinary driving, as illustrated in Figure 11.2. Where do these data come from? Technicians install recording equipment in vehicles that log thousands of miles, such as Uber and Lyft cars. These drivers go about their day, while cameras and sensors record videos of about everything they see. Everything includes moving and stationary objects, from pedestrians and cyclists to other cars, trucks, and moving vans, from traffic signals and trees to buildings, fences, and bridges. These examples, together with high-resolution digitized maps (think Google Maps on steroids), form the data for teaching the self-driving algorithm.

Drivers in Vehicles with
Sensors Recording Data

Recorded Miles with
Roadway Examples &
Highly Digitized Maps
Stored in the
Electronic Cloud

FIGURE 11.2   Process used to collect data needed to train a self-driving algorithm. Ordinary vehicles are equipped with sensors and recording devices to collect examples. The examples become the data used to train the self-driving system.

Next, each of the elements in the video footage needs to be identified and classified. Pedestrians, regardless of their height, shape, haircut, ball cap, or jacket, need to be clearly identified as humans. Different types of vehicles, from compact sedans and SUVs to pickup trucks, moving vans, bicycles, motorcycles, and emergency vehicles, receive a code. Everything in the video is reduced to an **object**.

In addition, the **action** of each object must be captured. The movement of a human stepping into a crosswalk, peddling a bicycle, or pushing a skateboard also is represented in the database. Similarly, the actions of the recorded vehicle, such as slowing or swerving around an object in the road or abruptly stopping when a ball rolls into the road, are all captured. Stationary-movable objects such as tree branches that could sway in the wind require special consideration. There also are stationary objects, such as signs, guardrails, or bridges.

If the self-driving algorithm can learn from these data how to correctly classify an object and predict its movement, then the system will know how to respond. As good drivers, we know to instinctively jam on the brakes when we see a ball roll into the road. Similarly, the algorithm needs to see the ball rolling, predict that a child may run after the ball, and quickly apply the brakes. The iterative, step-by-step process of training the algorithm is illustrated in Figure 11.3.

The volume of examples in these training sets is mind boggling. After all, an example also has contextual elements. Did the ball roll into the street from between parked cars? Or, did the ball first bounce, then roll into the street? Is a child likely to run into the street from a certain direction? Or, could a soccer player come running from either direction? Can the system recognize any object in dim light, in rain, or in windy conditions?

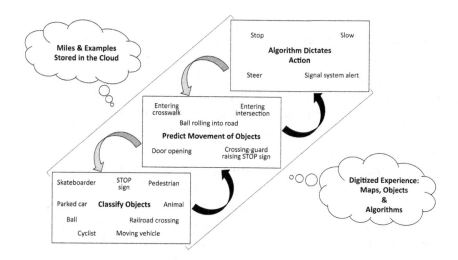

FIGURE 11.3 The iterative process of training the self-driving algorithm. The process uses the most advanced statistical and computer analysis tools, generally called deep learning, to build "experience" into the algorithm.

Recognizing the situation and knowing what to do next is at the heart of our driving experience. Rapidly apply the brake. Swerve the car. Our actions come from experience. The self-driving systems earn their experience using these data-intensive examples. The systems are trained on billions of examples using highly sophisticated statistical and computational algorithms. Sometimes called "deep learning," this training takes place in a research and development laboratory. The goal is to mimic the behaviors of the best drivers.

Once trained to recognize, predict, and make the right driving decisions, the computer system is ready for testing. The first series of tests also occurs in a laboratory. That is, the self-driving computer system is "let loose" within simulated-driving scenarios. You might think about this as a computer game, where an algorithm controls the joystick or lever that represents the steering wheel of a car. In this way, researchers test the self-driving system to make sure, for example, that it avoids obstacles, obeys traffic signals, and makes way for pedestrians.

In their February 2021 safety report, Waymo engineers reported that their system had completed over 15 billion miles of simulated driving. In June 2021, a Tesla engineer chose a different set of measurements to describe their volume of training. They had used 1 million videos of diverse scenarios with 6 billion objects to update their vision software. That is, lots and lots of data.

Once a company deems its computer system to be well trained in the laboratory, the next step is to try the system on a test track. The vehicle's cameras and sensors need to work with the self-driving algorithm in this test environment. The University of Michigan has one such track. The facility looks a bit like a sports arena, a totally enclosed, dedicated space. It has streets, traffic signals, buildings, trees, and other objects that represent a realistic driving environment. Logging miles on these test tracks adds to the algorithm's "understanding and experience."

These vitally important tests also are used to gauge the entire system's capabilities, including both its hardware and software. More realistic scenarios, different from the controlled examples seen in the laboratory, can task the system. For example, debris crosses the road. A street light goes out. A skateboarder suddenly enters an intersection.

Each time the self-driving vehicle is challenged, the researchers ask important questions. Did the system recognize the object on the road? Did it correctly predict the object's actions? Did it react appropriately? You might think of this as the driver's education phase of a self-driving vehicle.

Together, all these data, algorithms, training, and tests create the brain of the vehicle over time. Once it passes these tests, the system is deemed ready to be deployed on public roadways, as illustrated in Figure 11.4.

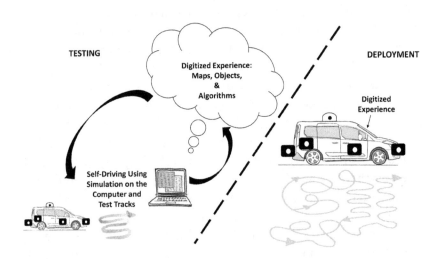

FIGURE 11.4 From testing to deployment. The self-driving system undergoes extensive testing before deployment to the public roads. The deployment uses test vehicles with the digitized experience from the training built into the vehicle's computer systems.

Waymo has one of the largest, on-the-road testing programs. As of February 2021, the company reported that its vehicles had logged over 20 million real-world miles on public roads in more than 10 states during the past decade. These real-world miles, together with more than 15 billion simulated miles, support the claim that Waymo's self-driving system is the World's Most Experienced Driver™.

THE READINESS DEBATE

Even so, some experts in the research community question the readiness of these systems to function independently on our roadways. They argue that computer systems are not as adaptive as humans because the algorithms cannot correctly respond to random, unforeseen, or slightly altered situations. They point to research that shows a computer system may not recognize a traffic signal partially obscured by tree branches, a plastic bag blowing across a roadway, or lane markings obscured by snow drifts. For example, one inter-university research team studied the problem of graffiti on traffic signs. Where a driver would ignore spray paint on a STOP sign, the team demonstrated that a self-driving algorithm misclassified a defaced STOP sign as a 45-mph sign.

Just as Tim Allen's character on the sitcom *Home Improvement* often would say, "need more power," engineers working in this field commonly respond to problems by saying, "need more data." That is, they usually attribute failures to a lack of complete training data. But since these systems already train with billions of objects and billions of miles, when will we know that they are safe enough to deploy?

Two investigators at the Rand Corporation, Drs. Nidhi Kalra and Susan Paddock, set out to address this issue. Based on their mathematical models and extensive analysis, they arrived at two key conclusions:

- "Autonomous vehicles would have to be driven **hundreds of millions of miles and sometimes hundreds of billions of miles** [emphasis added] to demonstrate their reliability in terms of fatalities and injuries."

- "Under even aggressive testing assumptions, existing fleets would take **tens and sometimes hundreds of years** [emphasis added] to drive these miles – an impossible proposition if the aim is to demonstrate their performance prior to releasing them on the roads for consumer use."

Clearly, this approach is not feasible. The Rand authors and others believe that there must be a more innovative or hybrid approach. Just as a human driver must pass an exam, the self-driving vehicle could demonstrate its capabilities by showing that it can safely navigate a reasonable set of real, on-the-road situations. Perhaps a mix of test scenarios could be used, including both simulated and actual miles.

But what scenarios should be included in such a set of tests cases? The snowy environment in Maine? The rain-drenched roads in Louisiana? Congested streets of New York City? The wide-open country roads in Oklahoma? In bright sunlight, in twilight, on well-lit roadways, or on dimly lit alleys? The possibilities are endless.

Certainly, the known failures could be used in the test cases. For example, a graffiti-marked STOP sign. Or, a reported Tesla problem in which the Autopilot mistook a low moon on the horizon for a yellow traffic light and slowed the vehicle. The key questions are what constitutes an appropriate, representative set of cases to qualify a self-driving vehicle, and what constitutes a test failure. This is an entirely new, unsolved field in automotive safety.

Still there are other, perhaps more complex, problems that need to be addressed to ensure safe deployment of these self-driving vehicles. After studying the Uber crash, the NTSB concluded that there were broader, system-related safety issues:

- Uber had an inadequate safety culture.

- The Arizona Department of Transportation fell short in its oversight of autonomous vehicle testing.

- The National Highway Traffic Safety Administration (NHTSA) was not doing its job in setting safety standards.

In other words, the NTSB found supervisory problems. This assessment is now part of the larger debate at the national level.

That debate includes other questions as shown in Figure 11.5. These range from federal versus states' rights to ownership of data. Between 2016 and spring 2021, the U.S. Department of Transportation and NHTSA published five reports addressing some of these questions. For instance, federal regulators deferred to the states to make their own decisions about allowing self-driving vehicles on their roadways. Also, the federal government encouraged the industry to voluntarily "prioritize safety." While

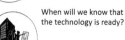

Who is in charge of assessing the readiness of the technology, the federal, or state government?

Which existing federal regulations still apply?

Are new rules and regulations need?

Who owns or has access to the data generated or used by self-driving vehicles?

When will we know that the technology is ready?

How detailed should laws be written to address cybersecurity threats?

In the event of a safety problem, who is responsible?

NHTSA

IIHS

CIVIC GROUPS

UNIVERSITY

LAW

NATIONAL ACADEMY OF SCIENCES

GOVERNMENT

INDUSTRY

FIGURE 11.5 National policy questions about self-driving vehicles. A wide cross-section of society has voiced concern about how self-driving vehicles will be regulated on our public roads.

this phrase is not clearly defined, the industry often points to highly reliable radars, cameras, and computers together with extensively tested algorithms to assert that safety is their highest priority. Nevertheless, these policies published under past presidents could change with future administrations.

Then there are the questions about data ownership and vulnerability. If a malfunction of the self-driving vehicle causes a crash, how will data be used to identify the problem? Who owns that data and who has access? We saw a glimpse of this problem in Chapter 9 when the Office of Defects Investigation had to rely upon Tesla to decipher the data associated with a possible sudden unintended acceleration problem. In that case, federal analysts were dependent on Tesla to assess the problem.

As vehicles use more electronics and software, there also is the problem of cyber threats. How vulnerable are these self-driving systems? And how are they tested to determine their ability to block a threat? After all, no one wants to experience a self-driving car run amok. Federal and local governments face a huge challenge to create appropriate policies that both promote progress and safeguard public safety.

While these big issues remain, nearly every local government has put in place some laws or executive actions to guide the deployment of self-driving vehicles on their roads. In many ways, our current challenges mirror the situation when horses and buggies shared the road with "mechanical horses." The entire safety community had to grapple with the problems of two vastly different modes of transportation, with different reaction times and ways of navigating. In those early days, states made decisions based on their perceived best practices.

Today, states appear to be following the same playbook. However, those decisions typically are made for small-scale fleets of test vehicles. For example, the Tweet mentioned at the beginning of this chapter shows that California permitted Cruise to deploy its test vehicles. This permission, though limited to a small fleet of vehicles in a confined, metropolitan area, is considered a breakthrough for the self-driving industry.

However, to move from limited testing in specific states to nationwide deployment, the larger questions shown in Figure 11.5 need addressing. In fact, the Infrastructure Investment and Jobs Act passed in November, 2021 directs agencies within the federal government to study these challenges. And the technology hardly matters. Whether the vehicles are passenger vehicles or long-haul trucks, whether these new vehicles drive on our streets or fly from rooftop to rooftop, the issues remain the same.

NOTE FROM THE AUTHOR

Data's role has changed over the history of car safety debates. In the past, society used data to identify and count the number of safety problems after the fact. Consumer advocates often employed one type of data analysis to call on the industry to design safer cars. Industry representatives countered using other data and types of analyses. As I described throughout this book, each side claims that the data, when correctly analyzed, support their position in the debate.

Today, data are being used for a wholly different purpose. Carmakers are using data to replace humans. The industry is moving data into the driver's seat, and betting that it will make our roads safer.

Will these robot-controlled cars be common in the near future? Will the industry's vision to save lives, injuries, and costs by replacing humans with machines be realized soon? According to the latest thinking by industry leaders, widespread deployment of self-driving vehicles in the United States is decades away. And with this long lead time comes an uncertain future.

Currently, driver-assist technologies are correcting many of the 94% of crashes attributed to "meathead" errors. Early data show that beeps, lights flashing, and emergency braking reduce crashes by as much as half. Over the years, as older cars retire and newer cars with these technologies become widespread, human drivers will avoid even more crashes.

As a result, we are confronted with an elephant-in-the-room-type question: Will driver-assist technologies remove the stated need for driverless cars?

In my opinion, the answer is yes. As our cars take on more robot-like features, crashes will become rarer. Fewer people will die or be injured in car crashes.

At the start of this chapter, I described the crash that killed Pamela Hesselbacher. Waymo engineers demonstrated that her young family would not have experienced a tragic loss if the pickup truck were self-driving. Perhaps the better question is, could driver-assist technologies, such as pedestrian detection or forward-collision warning, have saved Pamela's life?

I believe that the resources invested in creating the self-driving vehicle should have been and should be devoted to making our human-driven vehicles safer. If saving lives is truly the objective, then there are other, more direct ways to accomplish this goal than deploying robotic vehicles. Instead of replacing "meat" with "silicon," I suggest that the industry work to make human drivers better. Augment drivers with easy-to-use, cheaper, and smarter driver-assist technologies. Other countries are further ahead in making these technologies standard.

The regulatory bodies at the national level also need to address the myriad of issues highlighted in Figure 11.5. Either the nation finds a way to answer these important questions, or the industry will stall. A third alternative is that the industry will move on its own, and society will be left to address these issues ad hoc after a tragedy occurs.

In conclusion, I believe that the cost of owning a Level 5 self-driving vehicle in the near future will be too steep for the average car buyer. Owners of these vehicles will face a steady stream of fees for software updates. I predict the monthly membership plans will be like our mobile phones, but more expensive. Plus, insuring these vehicles and assigning blame in the event of a crash are huge unknowns. These uncertainties point to the conclusion that people like you and me are unlikely to be buying a Level 5 vehicle anytime soon.

Instead, I foresee self-driving vehicles becoming a new segment of the commercial transportation system, moving groups of people and goods across our roadways. These robots could be shuttles, taxis, delivery vehicles, and freight movers. In this way, the fleet of vehicles would be limited. A few companies would have ownership and the operators – who monitor these robots – could be certified. This commercial segment would have a specific market to serve. Once it is organized as a well-defined entity, then it could be regulated with nationwide, consistent rules. As a society, we already have experience in regulating something similar. We have our aerospace industry.

NOTE

1. California Public Utilities Commission. (2021, June 4). *CPUC approves first driverless autonomous vehicle service under pilot program designed to transform state's transportation system* [Tweet & Press release]. https://www. cpuc.ca.gov/news-and-updates/all-news/cpuc-approves-first-driverless-autonomous-vehicle-service-under-pilot-program.

Acknowledgments

Every career starts with an event that takes on meaning only after the passage of time. For me, it was a phone call from a colleague, Professor Art Dean at Arizona State University. He had turned down an opportunity to consult with a bioengineering firm about some motor vehicle crash statistics. He asked if I could take the project.

Shortly afterward, another safety consulting firm, Arndt & Associates, asked me to sort out some statistics problems. The National Highway Traffic Safety Administration (NHTSA) in a recall investigation had alleged that General Motors' (GM) C/K pickup trucks, with their side-saddle fuel tanks, posed a safety risk. Opposing this claim, GM engineers asserted that their vehicles were no worse than other similar pickups. Both organizations used data to support their positions. The rather open-ended question posed to me by Arndt & Associates: How could two opposing groups use the same data and arrive at two different conclusions?

That was 1991. Not much has changed. Since reading this book, hopefully you have an understanding that the devil is often in the details.

I am grateful to the many people over the past 30 years of my career. Early on, Attorney David Perry and the Honorable Renée Haas brought me many interesting problems in the form of legal complaints. Sadly, each started with a family's tragic story. My role has always been to understand and explain the real message behind the statistics. Most cases ended with a settlement and, in many instances, the manufacturers made design changes to their vehicles. Litigation is one way that car safety has advanced.

GM no longer builds side-saddle fuel tanks on their pickups. Jeep no longer builds rear-mounted fuel tanks on their SUVs. Attorneys David Perry and Patrick J. MacGroder III motivated Ford to change the rear-mounted fuel system design in their Crown Victoria Police Interceptor only after police officers were killed or wounded. I have worked with many talented firms including Langdon & Emison, Butler, Wooten & Peak, and The Gilbert Law Group. I am grateful to have been a part of these teams.

But none of this work could be done without the unsung heroes, the people bringing us the data. This book is dedicated to all the individuals – inside and outside of NHTSA – past, present, and future – who collect, code, and store crash data. From the frontline police officers, emergency medical teams, crash analysts, statisticians and data programmers to computer scientists, the fruits of their labor make it possible to identify and remedy car safety problems. I appreciate their daily dedication.

In particular, I want to acknowledge, Dr. Charles Kahane, for his persistent, high-quality research at NHTSA. His leadership in the application of statistical analysis to car safety cannot be overstated.

In addition, I want to recognize one organization and key individual at the University of Michigan Transportation Research Institute (UMTRI). In the early years, we could not easily download the massive, federal crash databases. Charlie Compton at UMTRI played an important role in providing access to the data and brainstorming about any analysis. He was insightful and light-hearted. It was a pleasure to work with him.

I crafted the chapters in this book to give you, the reader, a glimpse into some of the great car safety debates. However, there are dozens of other topics that I could have chosen. The literature is filled with thoughtful, scholarly papers reporting on the work occurring at the federal level and in universities, industrial and research organizations, consulting groups, and hospitals. My recognition and thanks go out to all the professionals working to save lives and reduce injuries by making our vehicles safer.

Many individuals influenced me during this writing journey. First of all, thanks to Professor Emerita Sharon Lohr, a friend and author, for encouraging me to write this book. Her guidance and positive attitude were tremendously meaningful. Professor Juliane Schober, also a friend and author, warned me that the book would take on a life of its own. She was right. I learned to be patient and allow the story to tell itself. Her advice was extremely helpful.

Many individuals have generously answered questions and helped to clarify some of the complex topics discussed in this book. I want to especially acknowledge Eric Teoh, Joe Nolan and Samuel Monfort of the Insurance Institute for Highway Safety for their research on incompatibility, Chuck Kahane for his work on fuel efficiency, and Nicholas Smith representing Waymo for the simulation work about autonomously driven cars. Attorneys Lynn Shumway and Brent Ghelfi were instrumental in showing me how families paid the price for the slow adoption of driver-assist technologies. A friend and car safety expert, Mark Arndt of Transportation

Safety Technologies, Inc., graciously read an early, partial manuscript and helped me to organize and scale some of the discussions. The book is better because of their wise advice.

Other gifted professionals lent me their expertise in finding vital information. Brad Vogus, Associate Liaison Librarian at Arizona State University, was expeditious in tracking down the records of congressional hearings and other critical federal documents. Also, I thank Sean Kane of Safety Research & Strategies for kindly sharing his extensive catalogue of materials on the incompatibility issue.

If I were to achieve my goal of reaching a broad audience, I needed to move away from my lifelong academic way of writing. Some very talented people especially helped me in this effort. Michelle Hubele Rubin of Your Writing Helper taught me that technical content could be woven into an interesting story for the non-technical reader. I may or may not have lived up to her vision, but I surely have a new way of looking at storytelling. And a new way to appreciate my daughter.

Henry Dembowski graciously read each chapter as it was developed. Henry used his skills as a teacher and coach to provide me with both broad and detailed feedback. He, together with his dear wife Claire, my cousin, played a key role in honing the message and keeping the project enjoyable. Also, Karen Murray Cady read the near-final manuscript. When she told me that a chapter "sang to her," I knew that I had hit the right note. I am truly grateful to these generous individuals.

The illustrations in this book were craftily drawn by Kendra Allenby. Her gifts of insight, drawing acumen, and humor made the writing task more pleasurable. And, I believe, the final product is more enjoyable to look at. Mucho gracias.

Laura Ingalls Fuqua of Abeja Solutions was invaluable in this project. Our braining storming sessions were essential for refining my message. Her editing and communication skills brought my ideas to life. One talented lady.

I cannot say enough about the insightful feedback from my two editors, first John Kimmel and then Lara Spieker of Chapman & Hall/CRC Press. They were very perceptive in leading me to a better presentation of these multifaceted topics.

This book complements the consumer information found on my website www.TheAutoProfessor.com. The vehicle safety ratings, called Auto Grades, are based on fatal crash statistics and are freely available on that site. That rating system and this book were heavily influenced by Dr. Katie Kennedy, who served as my chief technical officer. Her love of a good story, outstanding analytical skills, and unique perspective are true treasures in my life.

Closer to home, I appreciate my dear family – Dr. Ed Rubin, Donna Davies, Linda Bradenburg, Deb Nacewicz, Chuck Faris, and Joyce Leary. And I want to recognize my close friends – Dr. Anice Anderson, Silvia Arellano, Eileen Aufiero, Dr. Diana Calica, Andrea Cohen, Mike Edwards, Dr. Esma Gel, Jane Harrison, Kathy MacDonald, Cynthia McGurren and Sally Oscherwitz. Also, a special recognition goes out to readers of the earliest manuscript Mary Hake, Pam Longenecker and Carol Muggeridge. Together, they enthusiastically encouraged me to keep at it. For all their support, much-needed levity and good food, I am forever grateful. After all, one does not live on books alone.

And finally, my husband, Norman. Our lively discussions over the past 40 years have always spurred me to think deeper. No topic ever goes unchallenged, including those discussed in this book. His intellect, enthusiasm, and positive outlook inspires me on a daily basis. My appreciation for the life we share is unbounded.

Further Reading

CHAPTER 1: INTRODUCTION

1. Throughout this book, I use elements of real stories, such as Tommy's, to humanize and motivate the discussion. In most of these stories, I change the names, but retain the details of the crash, vehicles make, and model. However, I use real names in high-profile cases involving public figures, such as actors, and those that resulted in legislation.

2. The prophetic letter from the mother about her daughter's harrowing experience was written by Norma Jean Friend to Chrysler and was read into the trial transcript, Walden et al. v. Chrysler Group et al, Superior Court of Decatur County, State of Georgia, Case no. 12-CV-472, March 24, 2015, p. 303. The jury found for the plaintiff.

3. Details about the Jeep defect and recall investigation are provided in Chapter 9. In this chapter, I quote from the document closing the engineering analysis and resulting in the recall of some of the Jeep vehicles, see: https://static.nhtsa.gov/odi/inv/2012/INCLA-EA12005-9765.PDF

4. On January 16, 2021, Chrysler and Jeep came under the ownership of Stellantis, a global automotive company.

CHAPTER 2: A STRANGE START FOR A MOVEMENT

1. The DeHaven story and quotes come from his own description of his journey, see: DeHaven, H. (1970). Beginnings of crash injury research. In Brinkhous, K.M. (Ed.), *Accident pathology: Proceedings of an international conference, Sheraton Park Hotel, Washington, DC, June 6–8, 1968* (1st ed., pp. 8–11). National Highway Safety Bureau. https://www.google.com/books/edition/Proceedings/c-VWGjOwEdEC?hl=en&gbpv=1&dq=%22beginnings+of+crash+injury+research%22+dehaven&pg=PA8&printsec=frontcover

2. For a less personal take on DeHaven's life and accomplishments, see: Gangloff, A. (2013). Safety in accidents: Hugh DeHaven and the development of crash injury studies. *Technology and Culture*, 54(1), 40–61.

3. The original article was DeHaven, H. (1942). Mechanical analysis of survival in falls from heights of fifty to one hundred and fifty feet. *War Medicine*, 2, 586–596. It also was reprinted in *Injury Prevention*, 2000, 6(1), 62–68. For a more in-depth discussion of this paper and other articles of early thoughts on car safety, see: Haddon, W., Suchman, E., & Klein, D. (Eds.) (1964). *Accident research, methods and approaches*. Harper & Row.

4. For details about the ACIR's ejection study and the data used in Figure 2.2, see: Tourin, B. (1958). Ejection and automobile fatalities. *Public Health Reports*, 53(5), 381–391.

CHAPTER 3: CRASH DATA MAKE A DIFFERENCE

1. Details presented by Dr. Moore and the illustrations reproduce in Figures 3.1, 3.3, 3.4, and 3.5 can be found in *Investigation of highway traffic accidents, U.S. House of Representatives Subcommittee on Interstate and Foreign Commerce*, 84th Cong. (1956) (testimony of John O. Moore). Also, the quotes in the section "Others Weigh In" are from other testimonies that took place in this meeting.
2. Concerning Colonel John Stapp's title, he was lieutenant colonel at the time of his testimony.
3. President Lyndon B. Johnson's proposal for a cabinet-level, U.S. Department of Transportation that consolidated existing transportation agencies can be found at 89th Cong. Rec 4629 (1966) (message from Lyndon B. Johnson).

CHAPTER 4: MEASURING PROGRESS WITH DATA

1. The data used in Figures 4.1 through 4.6, 4.9 through 4.10 can be found in the Traffic Safety Facts Annual Report Tables at https://cdan.nhtsa.gov/tsftables/tsfar.htm
2. While FARS is highly regarded and widely used, it does have an important limitation. It does not record fatal crashes that occur in driveways and parking lots. For example, a special data collection project had to be undertaken to obtain counts of children tragically killed when backed-over in driveways. Similarly, the death of Star Trek actor Anton Yelchin is not recorded in FARS. His death was part of the controversy to recall Jeep Grand Cherokees due to an unconventional design of the parking gear. Yelchin allegedly was killed when his vehicle, assumed to be in park, rolled down his driveway and pinned him against a gate post.
3. The annual number of licensed drivers, also found in the Traffic Safety Facts Annual Report Tables, is obtained from individual states and may include some graduated, teen-licensed drivers.
4. For the suggestion that having only female drivers would halve the fatality rate, see the calculations at www.TheAutoProfessor.com/book
5. The number of registered vehicles, contained in the Traffic Safety Facts Annual Report Tables, is compiled by a private organization called R.L. Polk, a foundation of IHS Markit Automotive Solutions. The methodology for compiling these numbers changed in 2011. Therefore, prior registration figures cannot be directly compared. But for the purposes of this general discussion, the details in the different methods are not a major concern.
6. As cited in the Traffic Safety Facts Annual Report Tables, the annual number of miles driven is estimated by the Federal Highway Administration. The methodology for estimating number of miles traveled also changed in 2007. See: https://www.fhwa.dot.gov/policyinformation/travel_monitoring/tvt.cfm

7. For the self-reported data graphed in Figure 4.7, see: McGuckin, N., & Fucci, A. (2018, July). Summary of Travel Trends: 2017 National Household Travel Survey (Report No. FHWA-PL-18-019). Federal Highway Administration, p. 74.
8. For the early benefits of seat belts reported from Victoria, Australia, see: Trinca, G. W., & Dooley, B. J. (1977). The effects of seat belt legislation on road traffic injuries. *Australian New Zealand Journal of Surgery*, 47(2), 150–155.
9. The estimate of lives saved and the layout of Figure 4.8 is credited to Kahane, C. J. (2015, January). *Lives saved by vehicle safety technologies and associated Federal Motor Vehicle Safety Standards, 1960 to 2012 – Passenger cars and LTVs – With reviews of 26 FMVSS and the effectiveness of their associated safety technologies in reducing fatalities, injuries, and crashes* (Report No. DOT HS 812 069). National Highway Traffic Safety Administration. For the count of primary and secondary seat belt laws, see the website hosted by the Governors Highway Safety Association: https://www.ghsa.org/state-laws/issues/seat%20belts

CHAPTER 5: THE ROOF CRUSH RESISTANCE DEBATE

1. For a more technical version of this debate, see: Hubele, N. F., & Arndt, M. W. (2012). Vehicle safety standard update: A case study in a regulatory debate using statistical models. *Chance*, 25(4), 4–12. https://doi.org/10.1080/09332480.2012.752278
2. Notice of Proposed Rulemaking to Upgrade the Federal Motor Vehicle Safety Standard (FMVSS) No. 216, Roof Crush Resistance, was published in the Federal Register 70 FR 49223 on August 23, 2005. This document is the source for the supporting data for Figure 5.1. This also opened the docket NHTSA-2005-22143. The letter introducing the chapter appeared as document 68 in this docket. For discussion on the test procedures, see page 49231 of NHTSA-2005-22143-005.
3. The supplemental notice that extended the feedback period for the proposed upgrade of the standard appeared in the Federal Register (73 FR 5484). The docket associated with this supplemental notice is NHTSA-2008-0015. For more details about the NHTSA statistical study, see document NHTSA-2008-0015-0004: Strashny, A. (2007). *The role of vertical roof intrusion and post-crash headroom in predicting roof contact injuries to the head, neck, or face during FMVSS 216 rollovers* (Report No. DOT HS 810 847). National Highway Traffic Safety Administration. https://crashstats.nhtsa.dot.gov/Api/Public/ViewPublication/810847
4. For more information about the National Automotive Sampling System Crashworthiness Data System (NASS-CDS), see the Glossary.
5. The Figure 5.4 illustrating the test procedure was reproduced from "Figure 1 Test Device Orientation" of the Final Rule of FMVSS 216 Roof Crush Resistance standard published April 27, 1999, the in Federal Register (64 FR 22579).
6. The comments from the industry can be found in NHTSA's Final Regulatory Impact Analysis of the Final Rule in docket document NHTSA-2009-0093-004.

7. For more information about the series of Malibu tests with the restrained dummy, see: Bahling, G.S., Bundorf, R.T., Kaspzyk, G.S., Moffatt, E.A., Orlowski, K.F., & Stocke, J.E. (1990). *Rollover and drop tests – The influence of roof strength on injury mechanics using belted dummies. Society of Automotive Engineering* (SAE Technical Paper 902314). SAE International. https://doi.org/10.4271/902314. A critique and discussion of these tests can be found in the Australian authored article, Rechnitzer, G., Lane, J., McIntosh, A. S., & Scott, G. (1998). Serious neck injury in rollovers – Is roof crush a factor? *International Journal of Crashworthiness*, 3(3), 286–294. https://doi.org/10.1533/cras.1998.0076 or document 38 in the docket NHTSA-1999-5572.

8. For more information about the exchange between the industry consultants and the IIHS, see:

   a. NHTSA-2008-0015-0077 letter authored by Dr. Adrian Lund, President of IIHS. Also, Brumbelow, M.L., Teoh, E.R., Zuby, D.S., & McCartt, A.T. (2008). Roof strength and injury risk in rollover crashes. *Traffic Injury Prevention*, 10(3), 252–265. https://doi.org/10.1080/15389580902781343

   b. NHTSA-2008-0015-0081 letter authored by J. Padmanaban of JP Research, Inc. and E.A. Moffatt of Biomech, Inc.

   c. NHTSA-2008-0015-0088 letter authored by Dr. Adrian Lund, President of IIHS with extensive attachments.

   d. NHTSA-2008-0015-0091 letter authored by J. Padmanaban of JP Research, Inc. with extensive computer printouts.

9. For more information about the 2020 study, see: Kweon, K.-J. (2020). *Evaluation of FMVSS No. 216a, roof crush resistance, upgraded standard* (Report No. DOT HS 813 027). National Highway Traffic Safety Administration. https://crashstats.nhtsa.dot.gov/Api/Public/Publication/813027

CHAPTER 6: THE INCOMPATIBILITY DEBATE

1. For more information about the Peltzman effect and some discussion articles, see:

   a. Chong, A., & Restrepo, P. (2017). Regulatory protective measures and risky behavior: Evidence from ice hockey. *Journal of Public Economics*, 151, 1–11.

   b. Peltzman, S. (1975). The effects of automobile safety regulation. *Journal of Political Economy*, 83(4), 677–726.

   c. Sobel, R.S., & Nesbit, T.M. (2007). Automobile safety regulation and the incentive to drive recklessly: Evidence from NASCAR. *Southern Economic Journal*, 74(1), 71–84.

   d. Høye, A. (2019). Vehicle registration year, age, and weight – Untangling the effects on crash risk. *Accident Analysis and Prevention*, 123, 1–11. With the rise of SUV registrations in Norway, this article provides insight into the early injury trends.

   e. Hedlund, J. (2000). Risky business: Safety regulation, risk compensation, and individual behavior. *Injury Prevention*, 6, 82–90. This is the full text of the Haddon Memorial Lecture delivered at the Fifth World Conference on Injury Prevention and Control in New Delhi, India. The journal editor introduced the article as "the most complete, most perceptive, and well-balanced appraisal of this complex issue I have ever read." It is worth noting that Hedlund, while holding a high-ranking position at NHTSA, oversaw the implementation of the National Automotive Sampling and General Estimate Systems.

   f. For a broader understanding of when and where the "Gladiator Effect" is invoked, see: Acabchuk, R.L., & Johnson, B.T. (2017). Helmets in women's lacrosse: What the evidence shows. *Concussion (London, England)*, 2(2), CNC39. https://doi.org/10.2217/cnc-2017-0005. This work, supported by the National Institutes of Health, gives an overview of the arguments for and against requiring helmets. The "Gladiator Effect" suggests that helmets will make the female players more like the men, i.e., more aggressive.

2. For the complete article of the early work cited, see: Hollowell, W.T., & Gabler, H.C. (1996). NHTSA's vehicle aggressivity and compatibility research program. *Proceedings of the Fifteenth International Technical Conference on the Enhanced Safety of Vehicles*, 1, Paper No. 96-S4-O-01, 576–592. This article is the source of the data used in Figures 6.3 and 6.4.

3. Figure 6.5 is a reproduction of a graph in the report titled "Initiatives to Address Vehicle Compatibility" that appeared as NHTSA-2003-14623-0001 to open the docket. The "2003 Industry Commitment" is contained in the document NHTSA-2003-14623-0013.

4. As an example of an economic assessment of the incompatibility problem, see: Anderson, M.L., & Auffhammer, M. (2014). Pounds that kill: The external costs of vehicle weight. *Review of Economic Studies*, 81(2), 535–571.

5. The data in Figures 6.6 are derived from NHTSA Traffic Safety Facts Annual Report Tables, 1994–2019. The data in Figure 6.7 are derived from Passenger Vehicles (2020). *Traffic Safety Facts*. (Report No. DOT HS 812 962). National Highway Traffic Safety Administration. https://crashstats.nhtsa.dot.gov/Api/Public/ViewPublication/812962.pdf

6. For more information about the studies measuring compatibility and aggressivity after the industry commitment, see:

   a. Greenwell, N.K. (2012, May). *Evaluation of the Enhancing Vehicle-to-Vehicle Crash Compatibility Agreement: Effectiveness of the primary and secondary energy-absorbing structures on pickup trucks and SUVs* (Report No. DOT HS 811 621). National Highway Traffic Safety Administration. https://crashstats.nhtsa.dot.gov/Api/Public/Publication/811621.pdf

   b. Monfort, S.S., & Nolan, J.M. (2019). Trends in aggressivity and driver risk for cars, SUVs and pickups: incompatibility from 1989 to 2016. *Traffic Injury Prevention*, 20(sup1), S92–S96. This is the IIHS study discussed in the chapter.

7. For more information about the car buying trends, see reports by Edmunds (https://www.edmunds.com/industry/insights/) or Experian (https://www.experian.com/automotive/auto-market-trends-webinar-form).

8. For a satirical article about the industry's race to make bigger vehicles, see: Hubele, N. F. (2020, March 10). Open letter to Elon Musk and Mary Barra. *Medium*. https://medium.com/@theautoprofessortap/open-letter-to-elon-musk-and-mary-barra-636d2f20d9ab

CHAPTER 7: THE FUEL EFFICIENCY DEBATE

1. For a more intensive look at the data and methodology proposed by Dr. Leonard Evans, see:
   a. Evans, L. (1982, June 7–10). *Car mass and likelihood of occupant* fatality [Technical Paper No. 820807]. SAE International Passenger Car Meeting & Exposition, Troy, MI. https://doi.org/10.4271/820807
   b. Evans, L. (1984). Driver fatalities versus car mass using a new exposure approach. *Accident Analysis & Prevention*, 16(1), 19–36. https://doi.org/10.1016/0001-4575(84)90003-4
2. The data in Table 7.1 was published in the June 29, 1978 Federal Register (43 FR 28204).
3. The details of the credits and penalties are part of a complicated standard. Approximately every five years, NHTSA opens a new public docket and various parties debate the details of the proposed, new CAFE standards. These discussions can be very extensive. For instance, between 2018 and 2020 the docket-setting standards for model years 2021–2026 contained thousands of discussion papers. The final rule consisted of more than 1,100 densely written pages of text, charts, and graphs. In this chapter, the focus is on the statistical highlights of the car safety discussion that have influenced the CAFE standards over the past 20 years.
4. For an economic analysis of and details about the early thesis that the CAFE standards exacted a "safety tax," see: Crandall, R.W., & Graham, J.D. (1989). The effect of fuel economy standards on automobile safety. *The Journal of Law & Economics*, 32(1), 97–118.
5. The data used to graph Figure 7.3 were part of the justification included in the Final Rule for CAFE standards for model years 1987–88. See: Passenger Automobile Average Fuel Economy Standards for Model Years 1987–88, 51 C.F.R. § 35612 (1986). https://archives.federalregister.gov/issue_slice/1986/10/6/35579-35620.pdf#page=34. The caption for Figure 7.3 about the decrease in new car weight is based on data found in Table 2 of the above cited article by Crandall and Graham. The defense of this trend was voiced by Assistant Comptroller General Eleanor Chelimsky of the U.S. Government Accountability Office: *Automobile weight and safety: U.S. Senate Subcommittee on Consumers*, 102nd Cong. (1991) (testimony of Eleanor Chelimsky).
6. National Research Council. (1992). *Automotive fuel economy: How far should we go?* National Academy Press. https://rosap.ntl.bts.gov/view/dot/13034. The quote about the conflicting conclusions of statistical studies can be found on page 47.

7. National Research Council. (2002). *Effectiveness and impact of Corporate Average Fuel Economy (CAFE) standards.* The National Academies Press. https://doi.org/10.17226/10172. The two large blocks of text stating the committee's findings can be found on pages 2 and 77; the dissenters' comments can be found on page 117. The data used to produce the graph in Figure 7.4 was published as footnote #5 on page 27 of this report.

8. Throughout this chapter, the term weight refers to a vehicle's curb weight, which is the published weight of the vehicle. The vehicle is assumed to have a full fuel tank and other necessary fluids, but no occupants or cargo.

9. Dr. Charles J. Kahane led the analysis of the relationship between vehicle weight and fatality risk. For more information, see:

   a. Kahane, C. (1997, January 1). *Relationships between vehicle size and fatality risk in model year 1985–93 passenger cars and light trucks* (Report No. DOT HS 808 570). National Highway Traffic Safety Administration. https://rosap.ntl.bts.gov/view/dot/8727

   b. National Highway Traffic Safety Administration. (2010 March). *Final regulatory impact analysis, corporate average fuel economy for MY 2012-MY 2016 passenger cars and light trucks.* https://www.nhtsa.gov/staticfiles/rulemaking/pdf/cafe/CAFE_2012-2016_FRIA_04012010.pdf

   c. Kahane, C. (2012, August). *Relationships between fatality risk, mass and footprint in model years 2000–2007 passenger cars and LTVs* (Report No. DOT HS 811 665). National Highway Traffic Safety Administration. https://crashstats.nhtsa.dot.gov/Api/Public/ViewPublication/811665 It also appeared in Docket No. NHTSA-2010-0152-0023.

10. Dr. Puckett and Mr. Kindelberger followed Kahane's methodology closely. The three reports of 2016 are stand-alone reports, whereas the 2020 report is part of the Final Rule:

    a. Puckett, S.M., & Kindelberger, J.C. (2016, June). *Relationship between fatality risk, mass, and footprint in model year 2003–2010 passenger cars and LTVs – Preliminary report* (Docket No. NHTSA-2016-0068). National Highway Traffic Safety Administration. https://www.nhtsa.gov/sites/nhtsa.gov/files/2016-prelim-relationship-fatalityrisk-mass-footprint-2003-10.pdf

    b. Environmental Protection Agency. (2016, July). Draft technical assessment report: Midterm evaluation of light-duty vehicle greenhouse gas emission standards and corporate average fuel economy standards for model years 2022–2025. https://www.nhtsa.gov/staticfiles/rulemaking/pdf/cafe/Draft-TAR-Final.pdf

    c. The Safer Affordable Fuel-Efficient (SAFE) Vehicles Rule for Model Years 2021–2026 Passenger Cars and Light Trucks, Final Rule. 85 C.F.R. § 24174 (2020). The Trump administration changed the acronym from CAFE to SAFE. This document also contains the data on page 24744 for median weights graphed in Figure 7.9. https://www.federalregister.gov/documents/2020/04/30/2020-06967/the-safer-affordable-fuel-efficient-safe-vehicles-rule-for-model-years-2021-2026-passenger-cars-and

11. For more details about the feedback received by NHTSA on the early safety analysis, see: National Highway Traffic Safety Administration. (2010). Relationships Between Fatality Risk, Mass, and Footprint (Docket No. NHTSA-2010-0152). https://www.regulations.gov/docket/NHTSA-2010-0152

12. For monitoring the changes in the CAFE standards and associate statistical safety impact analysis, see: https://www.nhtsa.gov/laws-regulations/corporate-average-fuel-economy

CHAPTER 8: THE SAFETY RATINGS DEBATE

1. Joan Claybrook teamed with the Advocates for Highway and Auto Safety in writing the critique and vision for a way to improve the U.S. NCAP. Claybrook, J., & Advocates for Highway and Auto Safety. (2019, October 17). *NCAP at 40: Time to return to excellence.* Advocates for Highway and Auto Safety. https://saferoads.org/wp-content/uploads/2019/10/NCAP-at-40-Time-to-Return-to-Excellence-by-Joan-Claybrook.pdf. The report also is a good source for a listing of new car assessment programs worldwide. The press conference announcing this report included other recognized, international car safety advocates. For more details about Claybrook's career, see:

   a. Lemov, M.R. (2015). *Car safety wars: One hundred years of technology, politics, and death.* Fairleigh Dickinson University Press.

   b. Vinsel, L. (2019). *Moving violations: Automobiles, experts, and regulations in the United States.* Johns Hopkins University Press.

2. Consumers can visit the NHTSA ratings webpage (https://www.nhtsa.gov/ratings) to search the federal 5-star safety rating system for individual cars. NCAP history and details about the tests, including videos of cars crashing, also can be found there. The illustrations of the crash tests in this chapter are presented only to assist the reader to understand the general concepts desribed.

3. The quote about the impossibility of comparing safety ratings came from a letter written by NHTSA Chief Counsel Jonathan Morrison on October 17, 2018, in response to Tesla making "a number of misleading statements about the recent government 5-star safety ratings of the Tesla Model 3." In particular, Tesla promoted its blog (Model 3 achieves the lowest probability of injury of any vehicle ever tested by NHTSA | Tesla) on Twitter and tagged NHTSA saying, "There is no safer car in the world than a Tesla." Tesla NHTSA FOIA Response. (2019, August 6). PlainSite. Retrieved June 30, 2021, from https://www.plainsite.org/documents/fnrhg/tesla-nhtsa-foia-response/. For a satirical commentary about the exchange, see: Hubele, N. F. (2019, October 30). Is the Tesla Model 3 the superhero we need? *Medium.* https://medium.com/@theautoprofessortap/is-the-tesla-model-3-the-superhero-we-need-8b845387e877. The essay also can found on https://theautoprofessor.com/tesla-safety-rating-the-tesla-model-3/

4. The vehicle relative risk, or the overall injury risk derived from the tests in NCAP, is a weighted average. The weights used in this calculation are based on the injury patterns of belted occupants found in the

Crashworthiness Data System. These weights are based on model year vehicles dating back to 1999, and prior to the final decision of 2008. For details on these weights, the injury curves, and other features of NCAP, see: Consumer Information; New Car Assessment Program, 73 F.R. § 40016 (2008). https://www.federalregister.gov/documents/2008/07/11/E8-15620/consumer-information-new-car-assessment-program

5. The distribution of adult weights and heights is based on data collected during the years 2015–2018. These data were used to construct Figure 8.6. See: Fryar, C.D., Carroll, M.D., Gu, Q., Afful, J., & Ogden, C.L. (2021). *Anthropometric reference data for children and adults: United States, 2015–2018 (Report No. 46)*. National Center for Health Statistics. https://www.cdc.gov/nchs/data/series/sr_03/sr03-046-508.pdf

6. For more information about females and seniors in car crashes, with a view toward improving the dummies and injury curves used in car testing, see:

a. Gendered Innovations. (2021). *Inclusive crash test dummies: Rethinking standards and reference models.* http://genderedinnovations.stanford.edu/case-studies/crash.html#tabs-2

b. Linder, A. & Svedberg, W. (2019, June). Review of average sized male and female occupant models in European regulatory safety assessment tests and European laws: Gaps and bridging suggestions. *Accident Analysis and Prevention*, 127, 156–162. https://doi.org/10.1016/j.aap.2019.02.030. Also see: Linder, A. (2018, December). *Eva, the female crash test dummy* [Video]. TED Conferences. https://www.ted.com/talks/astrid_linder_eva_the_female_crash_test_dummy

c. Roberts, C.W., Forman, J.L., & Kerrigan, J.R. (2018, September 12–14). *Injury risk functions for 5th percentile females: ankle inversion and eversion* [Technical paper]. International Research Council on the Biomechanics of Injury Conference, Athens, Greece. http://www.ircobi.org/wordpress/downloads/irc18/pdf-files/103.pdf

d. Digges, K., Dalmotas, D., & Prasad, P. (2013, May 27–30). *An NCAP star rating system for older occupants* [Technical paper]. 23rd Enhanced Safety of Vehicles Conference, Seoul, Korea. https://www-esv.nhtsa.dot.gov/Proceedings/23/isv7/main.htm

e. Kahane, C. J. (2013, May). *Injury vulnerability and effectiveness of occupant protection technologies for older occupants and women* (Report No. DOT HS 811 766). National Highway Traffic Safety Administration.

f. Bose, B., Segui-Gomez, M., & Crandall, J.R. (2011). Vulnerability of female drivers involved in motor vehicle crashes: An analysis of US population at risk. *American Journal of Public Health*, 101, 2368–2373. (In the chapter, this was referred to as the University of Michigan study.)

g. Augenstein, J., Digges, K., Bahouth, G., Perdeck, E., Dalmotas, D., & Stratton, J. (2005). *Investigation of the performance of safety systems for protection of the elderly* [Technical paper]. Association for the Advancement of Automotive Medicine Annual Conference, Boston, Massachusetts, United States. https://www.ncbi.nlm.nih.gov/pmc/articles/PMC3217455/

7. The finding that males fitting the profile of the dummy used in NCAP testing had the fewest injuries in car crashes can be found in the article Carter, P.M., Flannagan, C.A., Reed, M.P., Cunningham, R.M., & Rupp, J.D. (2014). Comparing the effects of age, BMI and gender on severe injury (AIS 3+) in motor-vehicle crashes. *Accident Analysis & Prevention*, 72, 146–160. https:// doi.org/10.1016/j.aap.2014.05.024

8. For more details about the statistical methods to estimate the life-saving effectiveness of NCAP, see: Kahane, C. J. (2015, January). *Lives saved by vehicle safety technologies and associated Federal Motor Vehicle Safety Standards, 1960 to 2012 – Passenger cars and LTVs – with reviews of 26 FMVSS and the effectiveness of their associated safety technologies in reducing fatalities, injuries, and crashes* (Report No. DOT HS 812 069). National Highway Traffic Safety Administration. https://www-esv.nhtsa.dot.gov/ proceedings/24/files/24ESV-000291.PDF

9. For details about the current Insurance Institute for Highway Safety (IIHS) vehicle ratings, tests and criteria for achieving the *Top Safety Pick* award, see https://www.iihs.org/ratings. IIHS also publishes their test results for child safety seats and boosters.

10. For a summary of the IIHS safety testing program, see: Zuby, D. (2015, June 8–11). *Consumer safety information programs at IIHS* [Technical paper]. 24th International Technical Conference on the Enhanced Safety of Vehicles, Gothenburg, Sweden. https://www-esv.nhtsa.dot.gov/Proceedings/ 24/isv7/main.htm

11. Consumers can visit the Insurance Institute for Highway Safety *Driver death rates by make and model* webpage https://www.iihs.org/ratings/ driver-death-rates-by-make-and-model to obtain 2020 IIHS ratings based on driver fatalities.

12. Because I feel strongly that regular people need access to unbiased, real-life crash data now, my team of data scientists developed a free-of-charge rating system based on the federal Fatality Analysis Reporting System (FARS) data called Auto Grades (https://theautoprofessor.com/auto-grade-search). Unlike the 5-star safety rating and *Top Safety Picks* systems, Auto Grades gives car buyers the power to compare the safety ratings of small cars versus mid-size SUVs versus pickup trucks. In addition, people can further filter the Auto Grades based on the age and sex of the driver to choose the right vehicle for them.

CHAPTER 9: TWO RECALL DEBATES

1. For the annual number of vehicles affected by recalls graphically displayed in Figure 9.1, see: National Highway Traffic Safety Administration. (2021). *2020 Recall Annual Report*. https://www.nhtsa.gov/sites/nhtsa.gov/files/documents/ 2020_nhtsa_recall_annual_report_021021-tag.pdf

2. For the entire letter submitted by Senator Walter Mondale in support of recall provisions in the National Traffic and Motor Vehicle Safety Act of 1966, see: 112 Cong. Rec. 14247 (1966) (statement of Senator Walter Mondale). The 1974

act that required carmakers to cover the cost of repairing the defect is found at An Act to Amend the National Traffic and Motor Vehicle Safety Act of 1966, Pub. L. No. 93-492, 88 Stat. 1470 (1974). https://www.govinfo.gov/content/pkg/STATUTE-88/pdf/STATUTE-88-Pg1470.pdf

3. For a deeper look into the defect investigation process and the definition of a defect, see: National Highway Traffic Safety Administration. (2020). *Risk-based processes for safety defect analysis and management of recalls* (Report No. DOT HS 812 984). https://www.nhtsa.gov/document/risk-based-processes-safety-defect-analysis-and-management-recalls

4. The Vehicle Owner Questionnaire (VOQ) can be found at https://www.nhtsa.gov/report-a-safety-problem

5. The tragic story of Tommy's death is the same one described in Chapter 1. For more information and the source documents in the Jeep Grand Cherokee defect investigation, see NHTSA IDs DP09-005, PE10-031, and EA12-005. To access these documents, use the NHTSA recall portal, but instead of entering the Vehicle Identification Number, use the link "Search by NHTSA ID." National Highway Traffic Safety Administration. (n.d.). *Safety issues and recalls.* https://www.nhtsa.gov/recalls

6. For discussion of the difficulties of assessing defects in vehicles with extensive electronics, see: Transportation Research Board. (2012). *TRB special report 308: The safety challenge and promise of automotive electronics – Insights from unintended acceleration.* The National Academies Press. https://doi.org/10.17226/13342. The block of text about the difficulties arising from electronic-intensive automobiles can be found on page xii.

7 The definition of unintended acceleration is contained in footnote #1 in: National Highway Traffic Safety Administration. (2011, April 15). *Technical assessment of Toyota electronic throttle control (ETC) systems.* https://static.nhtsa.gov/odi/inv/2014/INRP-DP14003-61485.pdf

8. The source of the information in Table 9.1 and the documents associated with ODI's investigation into complaints about Tesla's alleged sudden unintended acceleration can be found in the investigation numbered DP20-001. Tesla's explanation about the VOQ #11206155 incident, as well as the reasons for denying the petition and closing the investigation can be found in Giuseppe, J.M. (2021, January 8). *ODI Resume: investigation DP 20-001.* National Highway Traffic Safety Administration. https://static.nhtsa.gov/odi/inv/2020/INCLA-DP20001-6158.PDF

9. In October 2021, an investigator with the Netherlands Forensic Institute published a paper about Tesla vehicle's logs: Hoogendijk, F. C. (2021). *Reverse engineering and evaluation of Tesla vehicle logs* [Technical paper]. 29th Annual Congress of the European Association for Accident Research, Haifa, Israel. These logs are important because they contain more information than the vehicle's electronic data recorder. They record data over the lifetime of the vehicle and store many more signals. The reverse engineering offers the opportunity to study these data independent of the Tesla corporation.

CHAPTER 10: THE AUTOMATED DRIVER-ASSISTANCE SYSTEMS DEBATE

1. Saving lives is the first entry in the mission of the Intelligent Transportation Society of America (itsa.org). This is a nationwide membership organization, including leadership from governmental transportation and planning agencies, private companies, auto manufacturers and suppliers, research organizations, academic institutions, and industry associations.

2. Treat, J.R., Tumbas, N.S., McDonald, S.T., Shinar, D., Hume, R.D., Mayer, R.E., Stansifer, R.L., & Castellan, N.J. (1977). *Tri-level study of the causes of traffic accidents: final report. Volume I: causal factor tabulations and assessments* (Contract No. DOT-HS-034-3-535). Institute for Research in Public Safety, Indiana University, Bloomington. https://babel.hathitrust.org/cgi/pt?id=mdp.39015075218019&view=1up&seq=7. For the results of NHTSA's 2005 study, see: National Highway Traffic Safety Administration. (2008). *National motor vehicle crash causation survey* (Report No. DOT HS 811 059). https://crashstats.nhtsa.dot.gov/Api/Public/ViewPublication/811059.

3. The list in Table 10.1 is compiled from a variety of sources, including the National Safety Council. (2021). *Car safety features. MyCarDoesWhat. org*. Retrieved September 13, 2021, from https://mycardoeswhat.org/safety-features

4. George Rashid's patent number 2,804,160 was granted on August 27, 1957. For the NHTSA report on his proposed safety system, see: Pollard, J.K. (1988). *Evaluation of the vehicle radar safety systems' Rashid radar safety brake collision warning system* (Report No. DOT-TSC-HS-802-PM-88-2). National Highway Traffic Safety Administration. https://rosap.ntl.bts.gov/view/dot/3127

5. For a detailed look at the early work estimating the benefits of driver-assist technologies and the data for Table 10.2, see: National Highway Traffic Safety Administration Benefits Working Group. (1996). *Preliminary assessment of crash avoidance systems benefits*. National Highway Traffic Safety Administration. In particular, the data reproduced in Table 10.2 is derived from page i and page 1–7 of this report.

6. The list in Table 10.3 is derived from Charles River Associates Incorporated (1998). *Consumer acceptance of automotive crash avoidance devices* (CRA Project No. 852-05). https://rosap.ntl.bts.gov/view/dot/2552

7. The report that chastised NHTSA for lack of action on driver-assist technology to prevent rear impacts is: National Transportation Safety Board. (2001). *Vehicle- and infrastructure-based technology for the prevention of rear-end collisions* (Special Investigation Report No. PB2001-917003, NTSB/SIR-01/01). https://www.ntsb.gov/safety/safety-studies/Documents/SIR0101.pdf. Among the crashes studies by the NTSB, three involved buses and one involved 24 vehicles.

8. The documents surrounding the industrial voluntary agreement to make automatic emergency braking standard can be found at: National Highway Traffic Safety Administration. (2015). *Automatic emergency*

braking initiative (Docket No. NHTSA-2015-0101). https://www.regula
tions.gov/docket/NHTSA-2015-0101

9. Figure 10.1 is based on the compilation of findings of the Insurance
   Institute for Highway Safety contained in their summary document *Real-
   world benefits of crash avoidance technology*, December 2020. https://
   www.iihs.org/media/259e5bbd-f859-42a7-bd54-3888f7a2d3ef/shuYZQ/
   Topics/ADVANCED%20DRIVER%20ASSISTANCE/IIHS-real-world-CA-
   benefits.pdf

10. For a survey of the literature on the effectiveness of a wide cross-section
    of ADAS, see: Tan, H., Zhao, F., Hao, H., & Liu, Z. (2021). Evidence of the
    crash avoidance effectiveness of intelligent and connect vehicle technolo-
    gies. *International Journal of Environmental Research and Public Health*, 18,
    9228. https://doi.org/10.3390/ijerph18179228

11. For the summary statistics about lives lost, injuries, and costs of motor
    vehicle crashes, see: Centers for Disease Control and Prevention. (n.d.). *Cost
    data and prevention policies*. Retrieved September, 20, 2021, from https://
    www.cdc.gov/transportationsafety/costs/index.html

CHAPTER 11: THE SELF-DRIVING CAR DEBATE

1. The opening quote is a Tweet by the California Public Utility Commission.
   For more on California permitting Cruise, LLC, to operate driverless
   vehicles, see: California Public Utilities Commission. (2021, June 4).
   *CPUC approves first driverless autonomous vehicle service under pilot pro-
   gram designed to transform state's transportation system* [Press release].
   https://www.cpuc.ca.gov/news-and-updates/all-news/cpuc-approves-
   first-driverless-autonomous-vehicle-service-under-pilot-program
   According to an article by National Public Radio, Cruise had logged
   over 2 million miles in California prior to its permit. Jones, D. (2021,
   June 5). California approves a pilot program for driverless rides.
   National Public Radio. https://www.npr.org/2021/06/05/1003623528/
   california-approves-pilot-program-for-driverless-rides

2. The fatal crash of Pamela Hesselbacher became a rallying cry for stronger
   sentencing of certain drivers in vehicular crashes. The driver of the pickup,
   William Epperlein, had a suspended license from a prior driving-under-
   the-influence conviction. He also neglected to carry the appropriate, high-
   risk insurance coverage. By law, the court could only sentence him to the
   maximum of 30 days in jail. Following lobbying efforts (https://pamsrights.
   com) by Pam's family, Arizona passed a new state law that imposes harsher
   penalties on such drivers who cause fatal or injurious harm in car crashes.

3. For more information about the Waymo simulation study see: Scanlon, J.,
   Kusano, K., Daniel, T., Alderson, C., Ogle, A., & Victor, T. (2021). Waymo
   simulated driving behavior in reconstructed fatal crashes within an autonomous vehicle operating domain. *Accident Analysis & Prevention*, 163,
   106454. https://doi.org/10.1016/j.aap.2021.106454

4. For a more technical discussion of the levels of automation, see: SAE International & ISO. (2021, April 30). *Taxonomy and Definitions for terms related to driving automation systems for on-road motor vehicles* (SAE J3016-202104). SAE International/ISO. https://www.sae.org/standards/content/j3016_202104/

5. For the 24-page report from the National Transportation Safety Board on the Tesla Model X crash, see: Karol, D., Becic, E., & Horak, D. (2018). *Automation and data summary factual report: Mt. View, CA* (No. HWY18FH011). National Transportation Safety Board. https://data.ntsb.gov/Docket/Document/docBLOB?ID=40480681&FileExtension=.PDF&FileName=Automation%20and%20Data%20Summary%20Report-Master.PDF

6. As I was finalizing the manuscript of this book, California prosecutors charged the first person in the United States with a felony for a fatal crash involving a Tesla on Autopilot. The driver, who worked for a limousine company, pled not guilty to two counts of vehicular manslaughter. While not the first criminal charges involving an autonomous vehicle, the case is likely to focus on the role of human error, the emerging technology, or both in the crash. For more details on the case, see: Krisher, T., & Dazio, S. (2022, January 18). *Felony charges are 1st in a fatal crash involving Autopilot.* Associated Press. Retrieved January 19, 2022, from https://apnews.com/article/tesla-autopilot-fatal-crash-charges-91b4a0341e07244f3f03051b5c2462ae

7. For more information about consumer behavior in Level 2 automated vehicles, see: State Farm Mutual Automobile Insurance Company. (2016). *Autonomous vehicles.* https://newsroom.statefarm.com/download/229160/2015autonomousvehiclesreport.pdf

8. For the report on the Uber crash, see: National Transportation Safety Board. (2019). *Collision between vehicle controlled by developmental automated driving system and pedestrian, Tempe, Arizona, March 18, 2018* (Highway Accident Report No. NTSB/HAR-19/03). https://www.ntsb.gov/investigations/AccidentReports/Reports/HAR1903.pdf

9. For an extensive and interesting description of Waymo's progress in building their fleet of autonomously driven vehicles, with an emphasis on safety, see: Waymo. (2021, February) *Waymo Safety Report.* https://waymo.com/safety/safety-report

10. For the multi-university study focused on classification problems, see: Eykholt, K., Evtimov, I., Fernandes, E., Li, B., Rahmati, A., Xiao, C., Prakash, A., Kohno, T., & Song, D. (2018). Robust physical-world attacks on deep learning visual classification. https://arxiv.org/pdf/1707.08945.pdf. This article is among numerous others referenced by the Duke University engineering and computer science professor Mary "Missy" Cummings in: Cummings, M. L. (2021). Rethinking the maturity of artificial intelligence in safety-critical settings. *AI Magazine* 42(1), 6–15. https://ojs.aaai.org/index.php/aimagazine/article/view/7394. In October, 2021, Cummings was named a new senior adviser for safety at NHTSA. Immediately after this announcement, Elon Musk of Tesla wrote on Twitter, "Objectively, her track record is

extremely biased against Tesla." For more information about the criticisms leveled by Tesla enthusiats and NTSB's chairperson Jennifer Homendy's response, see: https://www.cnn.com/2021/10/28/cars/tesla-ntsb-cummings/index.html. (Retrieved May, 25, 2022) Numerous media outlets reported in December 2021 that Cummings was removed as an expert witness for the plaintiff's side in a case involving the Tesla Autopliot, e.g., https://www.bloomberg.com/news/articles/2021-12-09/tesla-autopilot-critic-who-took-u-s-job-out-as-trial-witness. (Retrieved May, 25, 2022) For more information about Tesla fans' criticisms of Cummings, see: https://www.teslarati.com/tesla-missy-cummings-kicked-out-autopilot-trial/. (Retrieved May 25, 2022) According to several media outlets, NHTSA requires Cummings to recuse herself from all Tesla-related matters at the agency, e.g., see the Wall Street Journal article from January 18, 2022, https://www.wsj.com/articles/elon-musks-tesla-asked-law-firm-to-fire-associate-hired-from-sec-11642265007. (Retrieved May, 25, 2022)

11. For more details about the methodology used to estimate the number of miles needed to certify the safety of self-driving cars, see: Kalra, N., & Paddock, S. M. (2016). *How many miles of driving would it take to demonstrate autonomous vehicle reliability* [Technical Report]. RAND Corporation, Santa Monica, CA (pp. 1129–1134).

12. Government documents on guidelines for autonomous vehicles include:
    a. National Science & Technology Council & U.S. Department of Transportation. (2020). *Ensuring American leadership in automated vehicle technologies, automated vehicles 4.0.* https://www.transportation.gov/sites/dot.gov/files/2020-02/EnsuringAmericanLeadershipAVTech4.pdf
    b. Feig, P., Schatz, J., Labenski, V., & Leonhardt, T. (2019, June 10–13). *Assessment of technical requirements for level 3 and beyond automated driving systems based on naturalistic driving and accident data analysis* [Technical paper]. 26th International Technical Conference on the Enhanced Safety of Vehicles, Eindhoven, Netherlands. https://www-esv.nhtsa.dot.gov/Proceedings/26/26ESV-000280.pdf

13. For a comprehensive, but brief discussion of the policy issues surrounding autonomous vehicles, see: Canis, B. (2021, April 23). *Issues in autonomous vehicle testing and deployment* (Report No. R45985). Congressional Research Service. Retrieved September 7, 2021, from https://sgp.fas.org/crs/misc/R45985.pdf. The November, 2021 Infrastructure Investment and Jobs Act mandated that federal agencies continue to study the regulatory challenges posed by self-driving vehicles.

Glossary of Selected Terms

Automotive Crash Injury Research (ACIR): Co-founded by Hugh DeHaven at the Cornell University Medical College in 1953, this multidisciplinary group was organized to determine the cause of automobile injuries by crash investigations, to publish findings and to influence future vehicle designs to prevent injury and death. It was built on the success of Crash Injury Research Project developed for injury prevention in aircraft crashes.

Crash Injury Research & Engineering Network (CIREN): CIREN, a program within NHTSA, is a collaboration of medical and research organizations working to improve the prevention, treatment, and rehabilitation of motor vehicle crash injuries in order to reduce deaths, disabilities, and human and economic costs.

Crashworthiness: This is generally used to describe the protection provided by a motor vehicle in guarding against occupant injury or death in a crash.

Crashworthiness Data Systems (CDS): As a scientifically designed nationally representative, random sample of approximately 5,000 minor, serious, and fatal crashes, this federal annual collection effort was designed to provide detailed information for car safety research. It was managed by NHTSA within the National Automotive Sampling System (NASS). It was replaced in 2016 by the Crash Investigation Sampling System (CISS).

Curb Weight: It is the weight of the vehicle with standard equipment and all the fuels or other fluids used to operate the vehicle. No occupants or cargo are included in this weight number.

Fatality Analysis Reporting System (FARS): Started in 1975, FARS is a publicly available, nationwide census database of fatal crashes occurring on the nation's roadways.

Federal Motor Vehicle Safety Standards (FMVSS): These regulations, issued by NHTSA, are designed to implement laws passed by the U.S. Congress and fulfill the mission to prevent and reduce motor vehicle crashes and minimize deaths and injuries.

Gore: An area of land where two roadways either diverge or converge. The direction of traffic on both sides of the gore is the same. The area of the gore is bounded on two sides by the edges of the roadways, with a jointing point of divergence or convergence.

Insurance Institute for Highway Safety (IIHS): The organization is supported by auto insurers and insurance associations. The IIHS describes itself on its website as "an independent, nonprofit scientific and educational organization dedicated to reducing deaths, injuries and property damage from motor vehicles crashes through research and evaluation and through education of consumers, policy makers and safety professionals." The Highway Loss Data Institute is also listed on the website as sharing its mission.

Light Trucks or Light Truck Vehicle: These vehicles are built on a truck frame, including vans, minivans, panel trucks, pickups, and sport utility vehicle with a Gross Vehicle Weight Rating of 10,000 lb or less.

Most Harmful Event: As a key element in a crash database, this is the event within the sequence of events in a crash that resulted in the most severe injury. For example, a fire may have been the most harmful event because an occupant died from the fire and not the crash.

Motor Vehicle Crash: A motor vehicle crash is a transport crash that (1) involves a motor vehicle in-transport, (2) is not an aircraft accident or watercraft accident, and (3) does not include any harmful event involving a railway train in-transport prior to the involvement of a motor vehicle in-transport. In modern car safety literature, the term "accident" replaces the term "crash." Using the word crash allows for the possibility that the event had some preventable factors such as speeding, driver impairment, distraction, or vehicle malfunction. The term "accident" connotes an unpreventable event and is therefore disliked by some safety advocates. Nevertheless, the words accident and crash are often used interchangeably.

Motor Vehicle Fatality: This refers to a death that occurred within 30 days of a motor vehicle crash. Furthermore, the death was determined to have resulted from the crash.

National Highway Traffic Safety Administration (NHTSA): As part of the U.S. Department of Transportation, NHTSA is charged with setting and enforcing safety performance standards for motor vehicles and motor vehicle equipment. In addition, the **Office of Defects Investigation (ODI)** within NHTSA identifies safety defects and manages the recall process. NHTSA also works with State and local jurisdictions in the development and delivery of highway safety programs. Overall, its mission is to save lives, prevent injuries, and reduce the economic costs of road traffic crashes.

National Transportation Safety Board (NTSB): As an independent U.S. government investigative agency, its purpose is to determine the cause of certain civil, transportation-related crashes. The investigations discover lessons that can, in turn, be used to prevent future crashes.

Index

Note: Locators in *italics* represent figures and **bold** indicate tables in the text.

Printed in the United States
by Baker & Taylor Publisher Services